IEE CONTROL ENGINEERING SERIES 16

SERIES EDITORS: PROF. H. NICHOLSON
                    PROF. B.H. SWANICK

SYSTEMS MODELLING AND OPTIMIZATION

Previous volumes in this series:

Volume 1   Multivariable control theory
                J. M. Layton

Volume 2   Lift traffic analysis, design and control
                G. C. Barney and S. M. dos Santos

Volume 3   Transducers in digital systems
                G. A. Woolvet

Volume 4   Supervisory remote control systems
                R. E. Young

Volume 5   Structure of interconnected systems
                H. Nicholson

Volume 6   Power system control
                M. J. H. Sterling

Volume 7   Feedback and multivariable systems
                D. H. Owens

Volume 8   A history of control engineering, 1800–1930
                S. Bennett

Volume 9   Modern approaches to control system design
                N. Munro (Editor)

Volume 10   Control of time delay systems
                 J. E. Marshall

Volume 11   Biological systems, modelling and control
                 D. A. Linkens

Volume 12   Modelling of dynamical systems—1
                 H. Nicholson (Editor)

Volume 13   Modeling of dynamical systems—2
                 H. Nicholson (Editor)

Volume 14   Optimal relay and saturating control system synthesis
                 E. P. Ryan

Volume 15   Self-tuning and adaptive control: theory and application
                 C. J. Harris and S. A. Billings (Editors)

SYSTEMS MODELLING AND OPTIMIZATION

Edited by
Peter Nash
Control and Management Systems Division
University Enginering Department
Cambridge

PETER PEREGRINUS LTD
on behalf of the
Institution of Electrical Engineers

Published by: The Institution of Electrical Engineers, London and New York
Peter Peregrinus Ltd., Stevenage, UK, and New York

© 1981: Peter Peregrinus Ltd.

All rights reserved. No part of this publication may be reproduced, stored in a retrieval system or transmitted in any form or by any means—electronic, mechanical, photocopying, recording or otherwise—without the prior written permission of the publisher

British Library Cataloguing in Publication Data

Systems modelling and optimisation.
 - (IEE control engineering series; 16)
 1. Control theory
 I. Nash, Peter II. Series
 629.8'312   QA402.3

ISBN 0-906048-63-X

Printed in England by A. Wheaton & Co., Ltd., Exeter

Contents

List of contributors		ix
Preface		x
1	**Mathematical programming**	1
	P. Toint	
	1.1 Introduction	1
	1.2 Theoretical background	2
	1.2.1 First order conditions	5
	1.2.2 Second order conditions	8
	1.3 Algorithms for unconstrained optimization	11
	1.3.1 Algorithms without derivatives	13
	1.3.2 Algorithms that require first derivatives	13
	1.4 Nonlinear constrained optimization	21
	1.4.1 Primal methods	22
	1.4.2 Dual methods	23
	1.5 Some other mathematical programming problems	31
	1.5.1 Linear and quadratic problems	31
	1.5.2 Integer programming	31
	1.5.3 Non-differentiable programming	31
	References	33
2	**Dynamic optimization**	36
	C. Storey	
	2.1 Introduction	36
	2.2 The calculus of variations	36
	2.2.1 The Euler Equations	39
	2.2.2 Extensions to the simplest problem	43
	2.3 Optimal control	44
	2.3.1 The Minimum Principle	45
	2.3.2 Dynamic programming	47
	2.4 Computational methods	50
	2.4.1 Reduction to parametric form	50
	2.4.2 Gradient methods	54
	2.5 Other areas of research	56

		References	58
3	Decomposition methods in optimization G.W.T. White		61
	3.1	Introduction	61
	3.2	Theory of decentralized optimization	63
		3.2.1 A complex production system	63
		3.2.2 Primal coordination	66
		3.2.3 Dual coordination	69
	3.3	Conclusion	73
		References	75
4	Introduction to linear programming L.H. Campbell		77
	4.1	Introduction	77
	4.2	Form of a linear program	77
	4.3	Basic solutions	78
	4.4	The revised simplex method	80
	4.5	Convergence of the simplex algorithm	84
	4.6	Choosing an initial basic solution	86
	4.7	Post-optimality analysis	87
		4.7.1 Right hand side ranges	87
		4.7.2 Objective coefficient ranges	87
		4.7.3 Other changes	88
	4.8	Duality in linear programming	89
	4.9	Duality and the simplex method	92
	4.10	Operational considerations	93
		References	96
5	Decomposition in linear programming L.H. Campbell		97
	5.1	Introduction	97
	5.2	Block angular structure	98
		5.2.1 The Complete Master Problem	99
		5.2.2 Column generation	101
	5.3	The Decomposition Algorithm	102
	5.4	Decomposition in practice	103
		References	104
6	Real-time process optimization D. Foster		105
	6.1.	Introduction	105
	6.2	Real time optimization of an olefine plant	105
		6.2.1 An olefine plant at Wilton	107
		6.2.2 The computer control scheme	107
	6.3	Optimal operation of a small power station	111
		6.3.1 The optimization problem	113
7	Control theory in macroeconomics M.B. Zarrop		123
	7.1	Introduction	123

7.2	Macroeconomic models	124
7.3	Control via linear models	127
7.4	Perturbation models	129
7.5	Controlling a large model: I	130
7.6	Iterative respecification of the cost function	132
7.7	Controlling a large model: II	134
7.8	Closed loop v. open loop	136
7.9	Policy sensitivity	136
	References	139

8 Optimal operation of canal reservoirs — 141
P. Nash

8.1	Introduction	141
8.2	Modelling reservoir systems	142
	8.2.1 A simple optimal control model	144
	8.2.2 Optimal control on drawdown	145
	8.2.3 Optimal control on refill	147
	8.2.4 A two regime policy	148
8.3	Control of the Caldon Reservoirs	149
	8.3.1 Assessing the control rules	150
	8.3.2 Two strategical questions	151
8.4	Application to other systems	153
	References	155

9 Coal market modelling — 156
R.A. Blewitt

9.1	Introduction	156
9.2	Background	156
	9.2.1 Barnsley Area	157
	9.2.2 Previous market models	157
9.3	Construction of the market model	159
	9.3.1 Specifying the model	160
	9.3.2 Design of the model	161
	9.3.3 The pilot model	163
9.4	Conclusion	164

10 Modelling the spread of telecommunications in less developed countries — 166
G. Walsham

10.1	Introduction	166
10.2	Problems of telecommunications strategy	166
10.3	Role of the telephone in economic development	168
	10.3.1 The hierarchy of places	169
	10.3.2 Revenue per line and other factors	170
	10.3.3 Analysis of economic categories	171
	10.3.4 Further work	172
10.4	Strategic modelling	172
	10.4.1 General structures	173
	10.4.2 Model calibration	173
	10.4.3 Results from the model	174
	10.4.4 Further work	176
10.5	Comments and conclusions	176

11	**Model evaluation**	**178**
	J.M. Macieowski	
	11.1 Introduction	178
	11.2 What is a model?	179
	11.2.1 Behavioural models	180
	11.2.2 Decision models	181
	11.3 Static and dynamic behavioural models	183
	11.3.1 Static behavioural models	184
	11.3.2 Dynamic behavioural models	187
	11.3.3 Informal techniques	189
	11.4 Decision models	190
	11.5 Conclusions	196
	References	197
Index		**199**

List of contributors

P. Toint
Department of Mathematics, Facultes Universaires de Namur, Belgium

C. Storey
Department of Mathematics, University of Technology, Loughborough

G.W.T. White
Topexpress Ltd., Cambridge

L.H. Campbell
Durham University Business School

D. Foster
ICI Ltd., Wilton, Middlesborough

M.B. Zarrop
Control Systems Centre, UMIST, Manchester

P. Nash
Control & Management Systems Division, University Engineering Department, Cambridge

R. A. Blewitt
National Coal Board, Doncaster

G. Walsham
Control & Management Systems Division, University Engineering Department, Cambridge

J.M. Macieowski
Department of Engineering, Warwick University, Warwick

Preface

This book is drawn from material presented at a vacation school for research students in control engineering, held in Cambridge in July, 1980, under the sponsorship of the (then) Control Engineering Committee of the Science Research Council. The school is one of a programme of five, soon to be six, vacation schools, whose aim is to broaden the training of control engineering research students. This particular school aims to explore some of the areas where control engineering and operational research overlap.

That this overlap is extensive can readily be seen, just by skimming through the many journals devoted, under one name or another, to "systems modelling". Indeed, it seems to me that there are a very large number of problems and theoretical areas which defy categorization as specifically operational research or management science or control engineering: the monitoring and control of environmental pollution, the efficient utilization of scarce natural resources, planning and control of large systems, control of integrated manufacturing operations are all obvious examples. Moreover, the common ground in these two disciplines must surely grow, as using the ever greater power of ever smaller computers becomes an essential feature of both.

A complete survey of the common ground of control engineering and operational research is obviously impossible in either a one-week school or in book of the length of this one. The vacation school was designed to concentrate largely on problems to which optimization techniques could sensibly be applied, and this book is about such problems and related thoeretical areas almost exclusively. Chapters 1 and 2 are surveys, of static optimization (mathematical programming) and dynamic optimization respectively. Chapter 1 in particular examines the problems of efficiently solving a mathematical optimization problem, once posed, and includes extensive references to up-to-date research in the field of mathematical programming algorithms. Some of these methods re-appear in chapter 2, where the use of function-space analogues of finite-dimensional programming algorithms is one of the methods discussed for solving problems in optimal control.

Chapter 3 surveys the theory of optimization under constraints as it applies to large-scale, and particularly decentralized or hierarchical, systems. The theoretical advantages of different methods of coordination in such systems are discussed, and it is shown how the theory provides a useful insight into the operation of systems subject to control at different levels. A particular type of problem decomposition, that of Dantzig and Wolfe for linear programs, is discussed in more detail in chapter 5.

Linear programming itself is almost certainly the most widely used and successful optimization technique. Chapter 4 is devoted to an introduction to the subject, and includes, with simple examples, an explanation of how the central technique of linear programming - the simplex method - works. Unusually in an introductory treatment, this chapter includes a brief discussion of the operational considerations involved in implementing a linear programming solution.

Chapters 6,8 and 9 provide specific case studies of problems of management and control approached by optimization techniques. Chapter 6 is concerned with the optimal operation of parts of a petrochemical plant; chapter eight is concerned with the optimal operation of reservoirs on the U. K. canal system; Chapter 9 is about the use of linear programming in a model used for optimizing the marketing strategy of an area of the National Coal Board. Chapter 7 is also case study material, but of a more general nature, and examines the problems of applying the techniques of control theory, and in particular optimal control, to models of the economy.

As well as illustrating the application of theory discussed in the first five chapters, this case study material is intended to shed light on the problems of modelling systems in the ways implied by approaches involving optimization. Indeed, it is evident in some of the studies that the most important part of the modelling effort comes in setting up the model. Once this is done, solution can be relatively simple, and it is perhaps worth saying that, while the development of efficient general algorithms is clearly necessary and important, a surprisingly large number of optimization problems that arise in practice are solved in an ad hoc way. Indeed, it seems sensible always to look carefully at any optimization problem in case the solution to it, or something close to it, can be found by a combination of mathematical technique, and insight born of knowledge of the system being modelled.

At the other end of the scale from these considerations, it is to be remembered that considerable help can be given with the solution of what may present itself as a problem in optimization without attempting any formal optimization at all. Just to be able to construct a model which can elucidate the consequences of different options may be sufficient to enable a satisfactory course of action to be chosen. This is to some extent the point made by chapter 10, which describes some continuing work on the modelling of the growth of telecommunications systems in less developed countries. This chapter also illustrates the very first steps of building a model to help people to make decisions, when there is no clear framework yet established and the goals of planning are known at best vaguely.

All of the models discussed in the case studies have in common the feature that their essential output is recommendations for action, whether by a computer controlling a plant or an engineer controlling a reservoir system or a treasury minister concerned with the country's economy. If these recommendations are put into effect, there is a clear implication that the people who decide that they should be implemented 'believe' in the models. It is interesting to ask where that belief comes from, and whether there are any criteria by which a model can be measured to see whether it is credible. These are the questions addressed by the final chapter. Here the 'optimizing model' is analysed into the three components of behavioural, cost and solution sub-models, and methods for evaluating these – in particular the first two – discussed.

Editing this book has been both interesting and rewarding, and has been made much easier by the help of a number of people. I should like to express my thanks to all the contributors, as well as to the students who made the original vacation school so interesting. I have had help, advice and patience from Tim Hills of Peter Peregrinus, and considerable encouragement from Josie Spring of the Science and Engineering Research Council. I am very grateful to Dr. G.W.T. White, who in addition to making a contribution in the form of a chapter, spent a considerable amount of time modifying the software that was used in producing the book.

Peter Nash

Chapter 1
Mathematical programming
P. Toint

1.1 INTRODUCTION

A mathematical programming problem is usually of the following type. Associated with some situation under study, there exists a real valued function that measures the performance or quality of the system that is considered, and it is desired to modify the system so that this performance index or objective function is as small as possible. Usually it is not possible to modify a part of the system arbitrarily without regard to the others: there are constraints that link the different components of the system. The only allowable modifications are those which satisfy these constraints. One wishes to find a state of the system that gives to the objective function the least possible value, while satisfying the constraints.

This can be mathematically expressed in the following way. Assume that the state of the system under consideration can be adequately described by a vector of n real numbers, x say. Let the objective function be f(x) and suppose that the constraints are expressed by some equations and bounds involving the components of x. Then the problem can be formally stated as

P1: minimize $\quad f(x)$

$\quad$ subject to $\quad e_i(x) = 0 \quad , \quad i \varepsilon I_e \qquad (1.1a)$

$\qquad\qquad\qquad h_i(x) \geq 0 \quad , \quad i \varepsilon I_i \qquad (1.1b)$

$\qquad\qquad x \varepsilon \mathbb{R}^n$

where the (1.1a) and (1.1b) represent equations and inequalities that describe the constraints, and where I_e and I_i are sets of indices of the constrained functions.

It is clear that P1 is fairly general, and that further assumptions will be needed in order to obtain practically solvable problems. These assumptions will generate several overlapping classes of optimization problems, depending, for example, on the mathematical description of the objective function, and on the effective presence of constraints on the x variable. At the same time, the formulation of P1 does not include a number of practically important problems. For example, problems where the performance of a system has to be measured by more than one criterion cannot be readily put into this framework.

The generality of problem P1 has led to mathematical programming becoming a very wide area of research, and the variety of specific problems that are addressed as well as the variety of the proposed solutions make a complete discussion impossible. Consequently, we will restrict ourselves in this chapter to an exploratory survey, with the main emphasis on algorithms and without much in the way of proofs. We begin with some general theoretical background.

1.2 THEORETICAL BACKGROUND

Consider again the general problem P1, where we now assume that all the functions involved are everywhere twice continuously differentiable. The following paragraphs will deal with conditions on the derivatives of these functions which are necessary or sufficient for a vector x^* to be a solution. First, let us define what we mean by a solution.

Definition: The vector $x \in \mathbb{R}^n$ is feasible for P1 if x satisfies (1.1a) and (1.1b). The vector $x^* \in \mathbb{R}^n$ is a local solution of P1 if x^* is feasible and there exists a neighbourhood of x^* such that for all feasible x in this neighbourhood, $f(x) \geq f(x^*)$.

We choose to follow in this section the structure that is presented in the book by Fiacco and McCormick [13]. In order to proceed, we need another definition: that of the Lagrangian function associated with P1, namely

$$L(x,u,v) = f(x) - u^T e(x) - w^T h(x). \qquad (1.2)$$

Here e(x) and h(x) are vector-valued functions whose components are $\{e_i(x) : i \in I_e\}$ and $\{h_i(x) : i \in I_i\}$ respectively. In expression (1.2), the vectors u and w are called Lagrange multipliers or Lagrange parameters associated with problem P1. We shall discover that the Lagrangian function and its derivatives are of paramount importance in deriving optimality conditions.

We now state, without proof, a classical result, due to Farkas [11].

Lemma 1.1

Let $\{a^k: k = 0, 1, \ldots, q\}$ be a set of vectors in $\mathbb{R}^n$. If

$$z^T a^0 \geq 0$$

for every $z \in \mathbb{R}^n$ such that

$$z^T a^k \geq 0, \quad (k = 1, \ldots, q) \tag{1.3}$$

then there exist non-negative coefficients $\{\alpha_k: k = 1, 2, \ldots, q\}$ such that

$$a^0 = \sum_{i=1}^{q} \alpha_i \, a^i \tag{1.4}$$

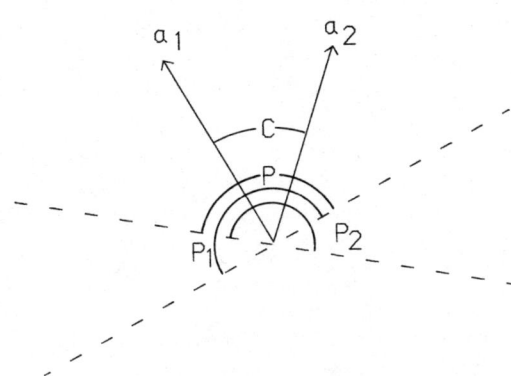

Fig 1.1 Farkas' Lemma.

This result is illustrated by the two-dimensional example in Fig 1.1. The sets of vectors satisfying (1.3) for $k = 1, 2$ are respectively P_1 and P_2. Hence the set of vectors satisfying both inequalities is P. It is easy to verify that the set of vectors that have non-negative projections on any vector of P is C, and hence that any such vector is a non-negative linear combination of a_1 and a_2.

Consider a feasible point x^* for P1 and define the following sets

$$I_i^* = \{i: i \varepsilon I_i, h_i(x^*) = 0\} \quad \text{(active set)}, \quad (1.6)$$

$$Z_1^* = \{z: z^T \nabla h_i(x^*) \geqslant 0 \text{ for } i \varepsilon I_i^*,$$
$$z^T \nabla e_i(x^*) = 0 \text{ for } i \varepsilon I_e, \ z^T \nabla f_i(x^*) \geqslant 0\}, \quad (1.7)$$

$$Z_2^* = \{z: z^T \nabla h_i(x^*) \geqslant 0 \text{ for } i \varepsilon I_i^*,$$
$$z^T \nabla e_i(x^*) = 0 \text{ for } i \varepsilon I_e, \ z^T \nabla f_i(x^*) < 0\}, \quad (1.8)$$

$$Z_3^* = \{z: z \varepsilon \mathbb{R}^n, \ z \notin Z_1^*, \ z \notin Z_2^*\} \quad (1.9)$$

The point of these definitions is that Z_1^* (Z_2^*) is the set of perturbations of x^* which, to first order, maintain feasibility with respect to equality and saturated inequality constraints and produce an increase (decrease) in objective function value. Z_3^* is the set of infeasible perturbations.

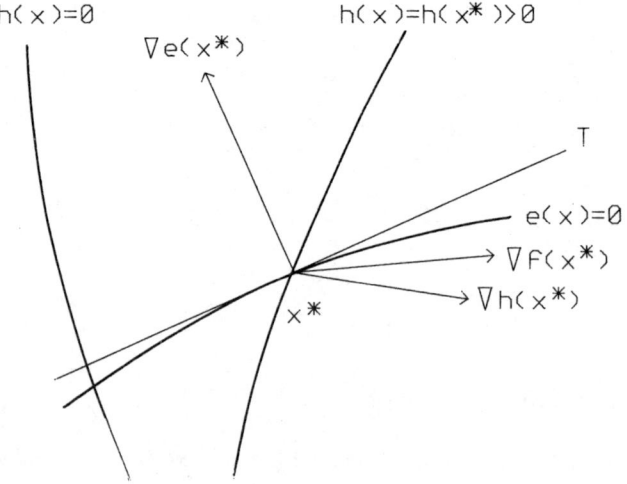

Fig 1.2 The sets Z_1^*, Z_2^*, Z_3^*.

Again let us consider a simple example (Fig 1.2). Observe that in this case x^* is certainly not a solution of the problem, since a move to the left on the curve $e(x)=0$ will decrease $f(x)$ while maintaining feasibility. This is reflected in the fact that

$$Z_2^* = \{z \varepsilon T: z^T \nabla f_i(x^*) < 0\} \tag{1.10}$$

is non-empty. Notice also that $\nabla h(x)$ does not play any role in (1.10) because the inequality constraint is not active: the requirement $z^T \nabla h_i(x^*) \geq 0$ in (1.7) and (1.8) is to maintain feasibility with respect to an already saturated inequality constraint.

1.2.1 First-order conditions

We now express a necessary condition depending on first-order derivatives in terms of x^* and Z_2^*.

Theorem 1.2

Let x^* be a local solution of P1 and assume that Z_2^* is empty. Then there exist vectors u^* and w^* such that

$$w_i^* h_i(x^*) = 0, \quad i \varepsilon I_i \tag{1.11}$$

$$w_i^* \geq 0, \quad i \varepsilon I_i \tag{1.12}$$

and

$$\nabla L(x^*, u^*, w^*) = 0. \tag{1.13}$$

That is

$$\nabla f(x^*) = \sum_{i \varepsilon I_e} u_i^* \nabla e_i(x^*) + \sum_{i \varepsilon I_i^*} w_i^* \nabla h_i(x^*). \tag{1.14}$$

Proof

By assumption Z_2^* is empty, so the vectors $\{\nabla h_i(x^*): i \varepsilon I_i^*\}$, $\{\nabla e_i(x^*): i \varepsilon I_e\}$, $\{-\nabla e_i(x^*): i \varepsilon I_e\}$ and $\nabla f(x^*)$ satisfy the conditions of Farkas' Lemma. Hence there exist non-negative coefficients $\{w_i^*: i \varepsilon I_i^+\}$, $\{y_i^*: i \varepsilon I_e\}$ and $\{v_i^*: i \varepsilon I_e\}$ such that

$$\nabla f(x^*) = \sum_{i \varepsilon I_i^*} w_i^* \nabla h_i(x^*) + \sum_{i \varepsilon I_e} (y_i^* - v_i^*) \nabla e_i(x^*). \tag{1.17}$$

Now, for $i \varepsilon I_e$, (1.14) is obtained by defining

$$u_i^* = y_i^* - v_i^* , \qquad (1.18)$$

while (1.12) is satisfied by construction and (1.11) results from the further definition

$$w_i^* = 0, \quad i \varepsilon I_i , \quad i \notin I_i^* . \qquad (1.19)$$

Several sets of conditions have been established which ensure that Z_2^* is empty, and consequently that (1.13) holds. One of the most useful has been stated by Mangasarian and Fromowitz [19] and is as follows.

<u>Theorem 1.3</u>

Assume that x^* is a local solution of P1 and that the following conditions are satisfied

(a) the vectors $\{\nabla e_i(x^*): i \varepsilon I_e\}$ are linearly independent;

(b) there exists a vector s such that

$$s^T \nabla e_i(x^*) = 0, \quad i \varepsilon I_e , \qquad (1.20)$$

$$s^T \nabla h_i(x^*) > 0, \quad i \varepsilon I_i^* . \qquad (1.21)$$

Then Z_2^* is empty, and consequently (1.11)-(1.14) hold.

The proof will not be given here and can be found in Fiacco and McCormick [13].

Conditions (a) and (b) of Theorem 1.3 are usually called "constraint qualification conditions". They imply that Z_2^* is empty. The converse is not true, as the following example, due to Kuhn and Tucker, shows.

$$\text{minimize} \quad x_2$$
$$\text{subject to} \quad h_1(x) = (1-x_1)^3 - x_2 \geq 0 , \qquad (1.22)$$
$$h_2(x) = x_1 \geq 0 ,$$
$$h_3(x) = x_2 \geq 0 .$$

This problem has the feasible region shown shaded in Fig 1.3.

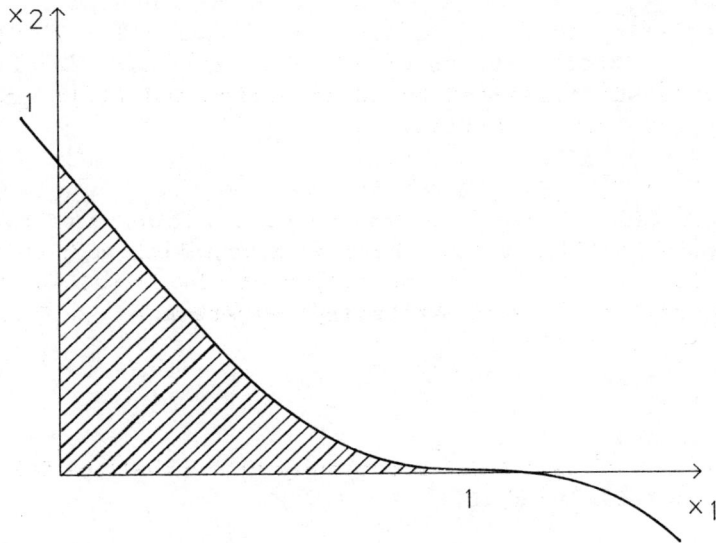

Fig 1.3 Problem for which constraint qualification is violated.

Consider the local solution $(1,0)$. The constraint qualification does not hold at this point because

$$\nabla h_1 = \begin{matrix} 0 \\ -1 \end{matrix}$$

(1.23)

$$\nabla h_3 = \begin{matrix} 0 \\ 1 \end{matrix},$$

and hence there is no vector that satisfies (1.21). On the other hand, Z_2^* is empty and there exist Lagrange multipliers

$$w = \begin{matrix} 0 \\ 0 \\ 1 \end{matrix}.$$

(1.24)

Observe that conditions (1.20) and (1.21) vanish if there is no inequality constraint, and then the only requirement of Theorem 1.3 is linear independence of the gradients of the equality constraints. If there are no constraints at all, then Theorem 1.2 reduces to the well-known condition that $\nabla f(x^*) = 0$.

To end this section on first order necessary conditions, we look briefly at the interpretation of the Lagrange multipliers, in the equality constrained case (see Powell [28]). Suppose that x^* is a local solution to P1 and that we perturb slightly the constraint (1.1a) to obtain

$$e_i(x) = \varepsilon_i, \quad i \varepsilon I_e . \tag{1.25}$$

Consider a feasible point of the perturbed problem of the form $x^* + z$, where $\|z\|$ is small. Then we have, for $i \varepsilon I_e$,

$$\varepsilon_i = e_i(x^* + z) = e_i(x^*) + z^T \nabla e_i(x^*) + O(\|z\|^2),$$

$$= z^T \nabla e_i(x^*) + O(\|z\|^2), \tag{1.26}$$

and we can deduce that

$$f(x^* + z) - f(x^*) = z^T \nabla f(x^*) + O(\|z\|^2)$$

$$= z^T (\sum_{i \varepsilon I_e} u_i \nabla e_i(x^*)) + O(\|z\|^2)$$

$$= \sum_{i \varepsilon I_e} u_i \varepsilon_i + O(\|z\|^2), \tag{1.27}$$

where we have used Theorem 1.2 in the equality constrained case. Thus, provided that the perturbed problem has a solution close to x^*, the Lagrange multipliers provide a first-order estimate of the change in optimal function value that results from a slight perturbation of the constraints. This is the reason why they are sometimes called shadow prices. Suppose that the constraints represent limitations on the use of some resources, and the objective function measures the benefit obtained by their use. Then the Lagrange multipliers represent the maximum prices worth paying for small increments of each resource. This point is discussed further in chapter 3.

1.2.2 Second-order conditions

We now outline some second-order conditions that can be used to distinguish between different extrema of P1. Consider the following example. Let P2 and P2' be two mathematical programming problems that differ only in their constraints:

P2: minimize $f(x) = x_1^2 + x_2^2$

subject to $x_2 - 1 \geq 0$

P2': minimize $f(x) = x_1^2 + x_2^2$

subject to $x_1^2 + x_2 - 1 \geq 0$

In Fig 1.4, the feasible set for P2 is above the horizontal line while that of P2' is above the parabola. The point $x^* = (0,1)^T$ is the solution of P2, and because the derivatives of both constraint functions are the same at this point, x^* also satisfies equation (1.13) for P2'. However, if we follow the parabola starting from x^*, we find that the function values decrease.

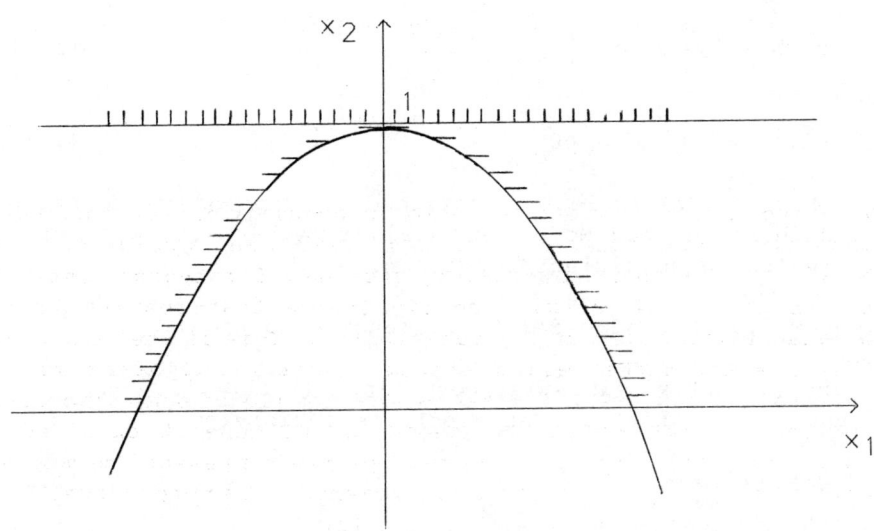

Fig 1.4 A stationary point which is not a solution.

Indeed, on the parabola,

$$f(x) = 1 - x_1^2 + x_1^4$$

which is less than $f(x^*)$ for small, non-zero x_1. This example illustrates two interesting points: a stationary point for the Lagrangian function is not necessarily a local solution, and this remains so even if in addition $\nabla^2 f(x^*)$, the second derivative matrix of the objective function, is positive definite.

Instead of considering $\nabla^2 f(x^*)$, one should consider

$$\nabla^2 L(x^*,u^*,w^*) = \{\frac{\partial^2 L}{\partial x_i \partial x_j}(x^*,u^*,w^*)\},$$

as the following results show.

Theorem 1.4 (Second Order Necessary Condition)

Suppose that x^* is a local solution of P1 and that the vectors $\{\nabla e_i(x^*): i \in I_e\}$ and $\{\nabla h_i(x^*): i \in I_i^*\}$ are linearly independent. Then

(a) Z_2^* is empty and (1.11) - (1.14) hold;

(b) if s satisfies condition (1.20) and in addition

$$s^T \nabla h_i(x^*) = 0, \quad i \in I_i^*, \tag{1.31}$$

then

$$s^T \nabla^2 L(x^*,u^*,w^*) s \geq 0 \quad . \tag{1.32}$$

The proof of (a) is left to the reader. A proof of (b) can be found in Fiacco and McCormick [13], together with a proof of the following sufficient condition.

Theorem 1.5 (Second-Order Sufficiency Condition).

Suppose that x^* satisfies conditions (1.11)-(1.14) for some u^* and w^*, and that for any vector s satisfying

$$s^T \nabla f(x^*) \leq 0, \tag{1.33}$$

$$s^T \nabla e_i(x^*) = 0, \quad i \in I_e, \tag{1.34}$$

$$s^T \nabla h_i(x^*) \geq 0, \quad i \in I_i^*, \tag{1.35}$$

it follows that
$$s^T \nabla^2 L(x^*,u^*,w^*) s > 0 \quad . \tag{1.36}$$

Then x^* is a local solution of P1.

Some remarks are now worth making:

(a) The set of vectors that is described by conditions (1.33)-(1.35) is slightly larger than Z_2^* since it includes <u>all</u> vectors that are orthogonal to $\nabla f(x^*)$.

(b) Neither Theorem 1.4 nor Theorem 1.5 require that $s^T \nabla^2 L s$ should be non-negative for all s in $\mathbb{R}^n$. It is quite possible that $\nabla^2 L(x^*, u^*, w^*)$ has some negative eigenvalues at a local solution of a constrained problem.

(c) If we consider the unconstrained problem the last remark is no longer true, since in that case conditions (1.32) and (1.36) reduce to positive semi-definiteness and positive definiteness of $\nabla^2 f(x^*)$.

(d) The linear independence condition in Theorem 1.4 can be somewhat weakened, but this leads to further complications (see Fiacco and McCormick [13] or Mangasarian [18]).

(e) Theorems 1.4 and 1.5 show clearly that, when solving a constrained optimization problem, the matrix $\nabla^2 L(x,u,w,)$ is much more important than the matrix $\nabla^2 f(x)$ because it incorporates information about the curvature of both the objective function and the constraints.

1.3 ALGORITHMS FOR UNCONSTRAINED NONLINEAR OPTIMIZATION

We now turn to the practical side and explore the unconstrained problem from the algorithmic point of view. Most of the currently available methods being based on the quadratic model, we first briefly describe this model.

One of the simplest (from the mathematical point of view) functions that admits a minimum in the absence of any constraint is the quadratic defined by

$$f(x) = \frac{1}{2} x^T A x + b^T x + c \tag{1.37}$$

where A is a positive definite symmetric n×n matrix, b is given vector in $\mathbb{R}^n$ and c is any real constant. It is clear that this function attains its global minimum at the point

$$x^* = -A^{-1} b \quad . \tag{1.38}$$

12 Mathematical programming

One idea that has been used extensively in the design of algorithms is that of generating, from a starting point x_0, a sequence of directions $\{s_i : i = 0, 1, \ldots, n-1\}$ so that when x_{i+1} is determined by a one-dimensional exact minimization on the line $x_i + \sigma s_i$, the relation

$$x_n = -A^{-1}b = x^* \tag{1.39}$$

is forced. One essential tool in achieving this purpose is the so-called quadratic termination property that is as follows.

Theorem 1.6

Let $\{s_i : i = 0, 1, \ldots, n-1\}$ be n vectors that satisfy the relations

$$s_i^T A s_j = 0, \quad i \neq j, \quad i, j = 0, \ldots, n-1. \tag{1.40}$$

Then the iterative procedure defined by

$$x_{i+1} = x_i + \hat{\sigma} s_i, \tag{1.41}$$

for given x_0, where $\hat{\sigma}$ is the solution to the problem

$$\underset{\sigma \in \mathbb{R}}{\text{minimize}} \; f(x_i + \sigma s_i) \tag{1.42}$$

and where f is as in (1.37), yields (1.39).

This is another classical result and we shall mention only that its proof relies heavily on the fact that $\hat{\sigma}$ is the exact solution to (1.42). Directions that satisfy (1.40) are called conjugate with respect to A.

Suppose now that the function to be minimized is not of the form (1.37). Taylor's Theorem implies that even so it can be locally approximated by such a quadratic, and one may therefore apply the techniques developed for (1.37) locally, and proceed in the hope the process will converge to the desired solution. Another basic observation that is fundamental to practical algorithms is the following. If one consider the point x and compute $\nabla f(x)$, then $-\nabla f(x)$ is the direction of steepest descent. One of the oldest algorithms (the steepest descent algorithm due to Cauchy) is based on this fact alone.

With these remarks in mind, we can separate practical algorithms into three classes, depending on the availability of the derivatives of the objective function $f(x)$.

1.3.1 Algorithms without derivatives

Suppose that the derivatives of f(x) are not available: the only information we have is function values. There are basically two ways to use these function values. One is to design an algorithm that evaluates the gradient $\nabla f(x)$ by differences, and the other is to use the function evaluations without trying to obtain information about derivatives.

The most successful algorithm of the second class is due to Powell [24], and has subsequently been improved by Zangwill [38] and Brent [6]. It is essentially an algorithm that builds conjugate directions on a quadratic objective function. This nice feature is possible because the line searches involved, i.e. problems of the form (1.42), are supposed to be exact. Although quite efficient, and very stable on problems with a moderate number of variables, its performance tends to deteriorate when the number of variables increases.

The algorithms of the second class are the same as those which require the gradient vector to be computed, with the supplementary difficulty that this vector must be estimated by differences along the coordinate directions. A way to obtain step sizes for this estimation has been designed by Stewart [35]. Practical algorithms built along these lines are also reasonably efficient. They do not require the line searches to be as accurate as those of the first class, but their efficiency also decreases as dimension grows because the gradient estimates become more and more expensive to obtain. Which approach is the best from the numerical point of view is not a clearcut question and research is still going on in this area.

1.3.2 Algorithms that require first derivatives

When the analytical gradient of the objective function is available, one has good information about the directions along which descent can be achieved. For this reason, the directions that are used to go from one iterate to the other usually involve the gradient vector explicitly. Again, two main classes exist: the quasi-Newton class and the conjugate gradient class (which includes the pure gradient method of steepest descent).

(a) Conjugate gradient algorithms

The first natural method that uses gradient information is the method of steepest descents, using the iteration described by

$$x_{i+1} = x_i - \hat{\sigma}\nabla f(x_i),\tag{1.43}$$

where again $\hat{\sigma}$ is chosen to solve (1.42), that is with exact line searches. This method is very simple to implement and can be proved to be convergent on a quite general class of functions, even if $\hat{\sigma}$ is not always determined exactly. However, it is computationally rather inefficient, mainly because of ill-conditioning of the objective function.

Consider a quadratic objective function of the form (1.37) and assume that the second derivative matrix A has eigenvalues of very different magnitudes. This leads to a function whose level contours look like Fig 1.5, forming a long narrow valley.

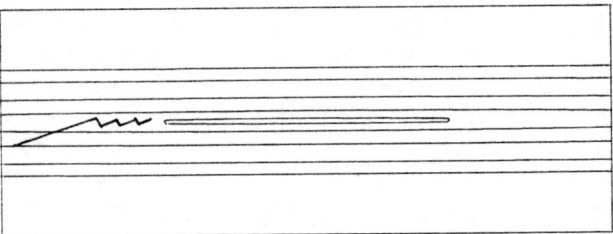

Fig 1.5 Zigzagging on an ill-conditioned objective.

Consider now the method defined by (1.43) and draw its trajectory on these level lines. We see that the steps become smaller and smaller, even far away from the optimum, so that they are eventually buried in the rounding errors of the computer and the algorithm makes no further progress. This zigzagging behaviour is typical of the method.

The presence of the narrow valley of Fig 1.5 may be considered as an intrinsic property of f(x). Indeed, the large difference of magnitude in the eigenvalues of A expresses the fact that the solution (1.38) is ill-conditioned with respect to A. As a measure of conditioning, we define k(A), the condition number of A, by

$$k(A) = \|A^{-1}\|_2 \|A\|_2 = \frac{\lambda_{max}(A)}{\lambda_{min}(A)},\tag{1.44}$$

where $\lambda_{min}(A)$ and $\lambda_{max}(A)$ are respectively the smallest and largest eigenvalues of A. The condition number of the function in Fig 1.5 is 50. One can also speak of ill-conditioning when the function, although not quadratic, presents deep narrow valleys in which zigzagging behaviour is most likely to take place. A nice example of such a function is

$$f(x_1, x_2) = \alpha \sin^2(\beta x_1) \cos^2(\beta x_2) + \frac{(x_1-x_2)^4}{1+(x_1-x_2)^4} + \frac{x_1^4}{1+x_1^4}. \quad (1.45)$$

Fig 1.6 is an isometric projection of part of the graph of this function near the origin, with $\alpha = 20$ and $\beta = 4$.

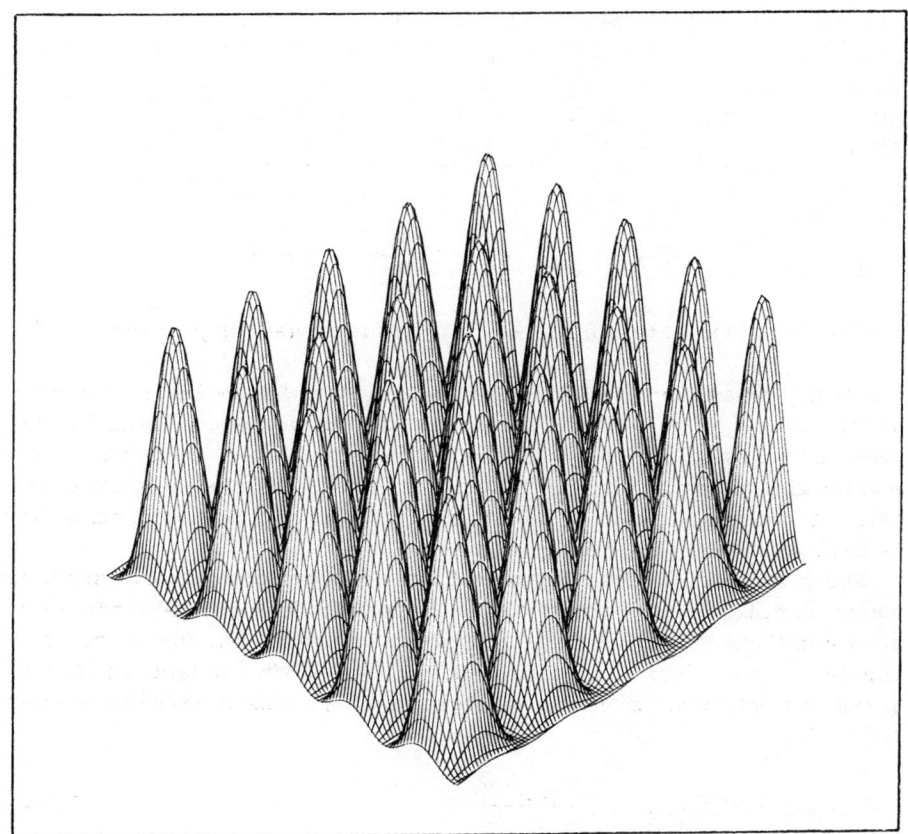

Fig 1.6 An ill-conditioned objective function.

Numerical experience indicates that most algorithms for unconstrained optimization suffer from this sort of ill-conditioning. Even if they do not stop far from the solution, like the gradient method, their progress is slower in deep, curved, narrow valleys.

To avoid the problem that appears in Fig 1.5, one may think of using the notion of conjugate directions together with the gradient information. A whole class of algorithms called conjugate gradients use an iteration of the form

$$s_i = -\nabla f(x_i) + \sum_{k=0}^{i-1} \beta_k s_k \quad , \tag{1.46}$$

$$x_{i+1} = x_i + \hat{\sigma} s_i \tag{1.47}$$

where (1.46) defines the new search direction at iteration i and where (1.47) yields the new iterate by an (exact) line search. The coefficients β_k in (1.46) are then determined so that the directions $\{s_i\}$ remain conjugate. If the function is quadratic and if the line searches are perfect, this implies that

$$\beta_k = 0, \quad k = 0, \ldots, i-2 \tag{1.48}$$

and therefore (1.46) becomes

$$s_i = -\nabla f(x_i) + \beta_{i-1} s_{i-1} \tag{1.49}$$

with β_{i-1} given by

$$\beta_{i-1} = \frac{\nabla f(x_i)^T (\nabla f(x_i) - \nabla f(x_{i-1}))}{s_{i-1}^T (\nabla f(x_i) - \nabla f(x_{i-1}))} \tag{1.50}$$

Relation (1.48) is important because it means that only the previous direction needs to be stored. Various methods have been designed to simplify expression (1.50) in the case of a quadratic objective function and exact line search. The most popular formulae are the Fletcher-Reeves [12] formula

$$\beta_{i-1} = \frac{\nabla f(x_i)^T \nabla f(x_i)}{\nabla f(x_{i-1})^T \nabla f(x_{i-1})} \tag{1.51}$$

and the Polak-Ribiere [23] form

$$\beta_{i-1} = \frac{\nabla f(x_i)^T (\nabla f(x_i) - \nabla f(x_{i-1}))}{\nabla f(x_{i-1})^T \nabla f(x_{i-1})} \quad . \tag{1.52}$$

Numerical experience favours the Polak-Ribiere form, and there are theoretical considerations which confirm this (Powell [25]).

We now consider some recent modifications of the conjugate gradient method. The first is concerned with the restarting strategy. In the classical algorithms of Fletcher-Reeves and Polak-Ribiere, s_0 was defined as $-\nabla f(x_0)$, and it has been observed that one can improve the convergence of these methods on non-quadratic objective functions by restarting the whole procedure after n iterations, that is at the point where the method could detect that the function was not quadratic. The question then arises: why throw out all the information we have obtained already by taking a new, pure gradient step? To overcome this problem, a conjugate gradient procedure of the form

$$s_i = -\nabla f(x_i) + \beta_{i-1} s_{i-1} + \gamma s_i \qquad (1.53)$$

where β_{i-1} is given by (50)-(52) and

$$\gamma = \frac{\nabla f(x_i)^T (\nabla f(x_1) - \nabla f(x_0))}{s_0^T (\nabla f(x_1) - \nabla f(x_0))} \qquad (1.54)$$

has been proposed by Beale [3]. This method allows the first search direction to be chosen as an arbitrary descent direction. (Other ways of allowing arbitrary initial search directions are due to Allwright [2] and Buckley [7]). Thus, when a restart is decided, one can simply start a procedure of the form (1.53)-(1.54) with s_0 the direction of the best step before the restart, and one thereby retains information from the previous cycle.

Related to this approach is the fact that the algorithm could detect the non-quadraticity of the objective function well before the nth iteration is reached. This is highly desirable, since when n is large, 1000 say, one does not want to wait for the 1000th iteration to decide to restart the procedure. An automatic restart device has been incorporated by Powell [25] in Beale's method (1.53)-(54) and this is one of the most successful routines available, especially when the dimension is large. (Note that we need to store only a few vectors, and no matrix at all, for all the conjugate gradient procedures.) A second way of extending the conjugate gradient methods is to modify them so that they do not depend too much on the exact line search. Propositions in this direction have been put forward by Dixon [10] and Nazareth [21].

Finally, a connection between the conjugate gradients and the quasi-Newton methods (studied in the next paragraph) has been demonstrated by Shanno [32] and used by Gill and Murray [14], but we leave this topic to the interested reader.

(b) Quasi-Newton methods

In contrast to the method studied above, requiring the storage of only a few vectors, we now turn to a class of methods that require an n×n symmetric matrix to be stored in memory. Consider again the quadratic model (1.37) and equation (1.38). This may be rewritten as

$$x^* = x - A^{-1} \nabla f(x_0),$$
$$= x - A^{-1}(Ax + b), \qquad (1.55)$$

for any given x in $\mathbb{R}^n$. Unfortunately, using (1.55) to generate x^* from x implies that $A = \nabla^2 f(x)$ is known, which is not the case. However, we still may think that an iteration of the form

$$s_i = -B_i^{-1} \nabla f(x_i), \qquad (1.56)$$
$$x_{i+1} = x_i + \hat{\sigma} s_i \qquad (1.57)$$

has some virtues, especially if we choose the matrix B_i to incorporate available information about the second derivative matrix of the objective function. A possible way of forcing B_i to approximate $\nabla^2 f(x_i)$ in some sense is to ask that B_i satisfies relations that are satisfied by $\nabla^2 f(x_i)$ when f(x) is quadratic, or by some matrix that is close to it when f(x) is non-quadratic. The best-known conditions are

$$B_i = B_i^T, \qquad (1.58)$$
$$B_i(x_i - x_{i-1}) = \nabla f(x_i) - \nabla f(x_{i-1}) \qquad (1.59)$$

where (1.58) holds for $\nabla^2 f(x_i)$, and the Quasi-Newton equation (1.59) holds with B_i replaced by

$$\int_0^1 \nabla^2 f(x_{i-1} + t(x_i - x_{i-1})) dt . \qquad (1.60)$$

However, (1.58) and (1.59) do not determine B_i uniquely, and one must design a way to take up the remaining degrees of freedom.

Most of the successful methods achieve this by choosing B_i to minimize

$$\|B_{i-1} - B_i\| \tag{1.61}$$

for some particular matrix norm. This is based on the fact that B_{i-1} will already contain valuable information about the second derivatives and the curvature of $f(x)$.

Another possibility is to compute directly B_i^{-1} instead of B_i, so that the computation involved in (1.56) is shortened. We now impose condition (1.58), and replace (1.59) by

$$B_i^{-1}(\nabla f(x_i) - \nabla f(x_{i-1})) = x_i - x_{i-1} . \tag{1.62}$$

If one now chooses B_i^{-1} to minimize

$$\|B_{i-1}^{-1} - B_i^{-1}\| \tag{1.63}$$

instead of (1.61), one obtains another class of possible methods for computing x_{i+1} via (1.56) and (1.57).

Probably the most successful method of this form is the well known Broyden-Fletcher-Goldfarb-Shanno method that uses the norm defined by

$$\|X\|^2 = \text{Tr}(B_i X B_i X^T) \tag{1.64}$$

(a weighted Euclidean norm) in (1.63) and results in the formulae

$$B_0 = I$$

$$B_i = B_{i-1} + \frac{y_i y_i^T}{d_i^T y_i} - \frac{B_{i-1} d_i d_i^T B_{i-1}}{d_i^T B_{i-1} d_i} \tag{1.65}$$

where

$$y_i = \nabla f(x_i) - \nabla f(x_{i-1}) , \tag{1.66}$$

$$d_i = x_i - x_{i-1} . \tag{1.67}$$

The expression (1.64) is indeed a norm, because each of the matrices $\{B_i : i = 0, 1, 2, \ldots\}$ is positive definite. In terms of B_i^{-1}, (1.65) becomes

$$B_{i+1}^{-1} = B_i^{-1} + \left\{ \frac{\|y_i\|^2 + d_i^T y_i}{(d_i^T y_i)^2} \right\} d_i d_i^T - \frac{d_i y_i^T + y_i d_i^T}{d_i^T y_i} , \tag{1.68}$$

after using the Sherman-Morrison inversion formula. This derivation of minimization methods by variational means is well motivated and the properties of such methods are described in the very good survey paper by Dennis and Moré [9].

Another point of view for generating methods of the type (1.56)-(1.57) is to ask that instead of (1.61) or (1.63), the method yield the optimum of a quadratic after at most n steps (with exact line searches). This point of view has been taken by Huang [17] and others and generates a wide class of updating methods which overlaps the class just described. For example, method (1.65)-(1.68) belongs to both classes. Another computationally prominent member of this class is a method due to Davidon [9] which is designed to take into account the conditioning of the matrices involved. Other methods that take conditioning into account have been proposed by Oren [22] and Shanno [31].

An interesting feature of the quasi-Newton methods, i.e. methods of the form (1.56)-(1.59), is that they rely much less on the precision of the computations in the line searches of (1.57). One can even prove (under certain assumptions) that the value $\hat{\sigma}=1$ becomes asymptotically exact. The line search therefore helps convergence from far away starting points but is of little importance in the asymptotic speed of convergence of the procedure. This is not true at all for the conjugate gradient type methods.

Finally, we mention a new direction of research for quasi-Newton methods: the design of procedures and formulae that take structure and sparsity into account. One says that an unconstrained minimization problem has a sparse structure whenever the second derivative matrix $\nabla^2 f(x)$ has a large proportion or interesting distribution of zero entries. This happens quite often in large-dimensional problems, especially those arising from the discretization of continuous ones.

Suitable methods have been proposed to take advantage of this structure by Toint [36] and [37], Shanno [33] and Marwill [20]. They are derived in the way described by (1.56)-(1.61) with the introduction of the supplementary (to (1.58) and (1.59)) constraint

$$(B_i)_{jk} = 0, \quad (j,k) \in I, \qquad (1.69)$$

where I is a set of given pairs of integers that describes the sparsity pattern of $\nabla^2 f(x)$. Their main advantages are reduction of the storage requirements and improved speed of convergence. A survey of methods of this type has been given by Toint [37].

An alternative method for taking the sparsity of $\nabla^2 f(x)$ into account has been proposed by Powell and Toint [27]: one uses differences in the gradient vector to estimate all the non-zero entries of the Hessian. Sparsity and symmetry properties are used to reduce substantially the number of differences that have to be calculated, so that the overall method is computationally competitive.

(c) Methods that require second derivatives of $f(x)$

Most of the methods that use an analytical computation of the Hessian matrix $\nabla^2 f(x)$ are based on Newton's algorithm, described by

$$x_{i+1} = x_i - \nabla^2 f(x_i)^{-1} \nabla f(x_i) , \qquad (1.70)$$

which is an extension of (1.55) to the non-quadratic case. Modifications of this method are mainly concerned with the problem of singularity and conditioning of the Hessian matrix and on extending the convergence region of iteration (1.70) to a wider area.

However, it is often the case that an analytical computation of the n^2 elements of $\nabla^2 f(x)$ is too expensive or too lengthy to be realistic as a practical procedure. This is the reason why such methods are seldom used, except when the Hessian is obtained directly as an extra product of other computations.

1.4 NONLINEAR CONSTRAINED OPTIMIZATION

We now return to the general problem of mathematical programming, P1. All the procedures that attempt to solve this problem numerically are of an iterative nature, as in the previous section. However, the presence of constraints introduces one more distinction: are all the iterates feasible, or does the method consider points that do not satisfy the constraints? In the next paragraphs, we will consider some methods of both types, with an emphasis on methods that allow non-feasible points to be considered, called dual methods, in opposition to primal methods that consider only feasible iterates. We will assume that first derivatives of both objective function and constraints are available for an analytical computation, since most of the practical algorithms for constrained minimization rely on this feature.

1.4.1 Primal Methods

As noted above, the main idea of the primal methods is to find the next iterate by a move that maintains feasibility. These types of method are sometimes called "admissible directions" methods. For the purpose of illustration, assume that in P1 there is no inequality constraint and that all the equality constraints are linear in x. A typical situation is described in Fig 1.7.

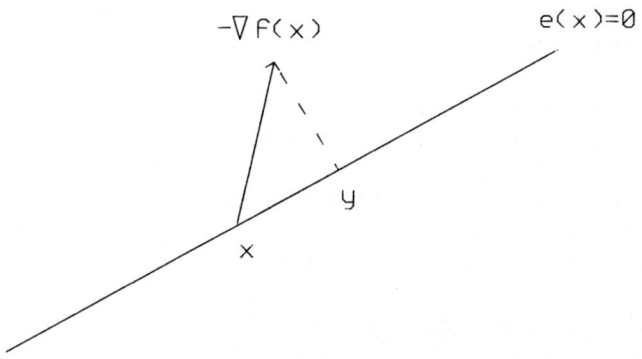

Fig 1.7 A step that maintains feasibility.

In this situation, it is clear that we can improve on f(x) by moving to the right, along the constraint. The direction that seems appropriate for searching for the next iterate is therefore y, the projection of $-\nabla f(x)$ on the feasible set. Notice that we used the descent direction $-\nabla f(x)$ in this example, but that any descent direction not orthogonal to the constraint is suitable. The problem becomes more complex when the constraints are nonlinear and when the problem includes inequalities or bounds on the variables.

A number of algorithms have been proposed that include various projection and feasibility enforcement steps. Amongst the most successful is the Generalized Reduced Gradient of Abadie [1] which is used in solving a number of real-life problems. Methods of this type that apply only for linear constraints have been proposed by Goldfarb [15] and Ritter [29]. Recently, Shanno and Marsten [34] have proposed a method that uses the projections onto the constraints of conjugate gradient directions. Because of the technicalities inherent in these approaches, we now turn to the dual procedures. However, we should to point out that primal methods are numerically quite successful.

1.4.2 Dual Methods

In these procedures, we allow our current iterate to be anywhere in $\mathbb{R}^n$ and only ask that the limit point that is produced by the iterations does satisfy the constraints. We also wish to use efficient unconstrained minimization techniques and eventually solve a sequence of unconstrained problems whose solutions tend to the solution of our constrained problem.

(a) Penalty approach

In this approach, one adds to the objective function another function that penalizes constraint violation. Thus one considers functions of the type

$$P_r(x) = f(x) + \sum_{i \in I_e} \frac{e_i^2(x)}{r_i} + \sum_{i \in I_i} \frac{1}{r_i} [\min\{0, h_i(x)\}]^2 \quad (1.71)$$

where the scalars $\{r_i\}$ control the amount of penalty for constraint violation that is introduced in the function $P_r(x)$. It is then possible to prove that the minima $\{x_r^*\}$ of the functions $\{P_r(x)\}$, which are the solutions to a sequence of unconstrained minimization problems, tend to x^*, the constrained optimum, as the factors $\{r_i\}$ tend to zero. This is to say that higher and higher penalties give better and better approximations of the solution of the constrained problem.

This basic approach, sometimes called SUMT, for Sequential Unconstrained Minimization Technique, has been investigated by Fiacco and McCormick [13] and others. Unfortunately, there is a major numerical problem that arises naturally with these types of calculations. For example, consider the very simple program

$$P3: \text{minimize} \quad (x^2 + y^2) \quad (1.72)$$

$$\text{subject to} \quad y + x^2(x^2 - 3) - 1 = 0 \quad (1.73)$$

whose solution is given by

$$x^* = 0, \quad y^* = 1, \qquad (1.74)$$

as can be verified by a straightforward calculation. The contour lines of the objective function and the curve representing the feasible set are shown in Fig 1.8.

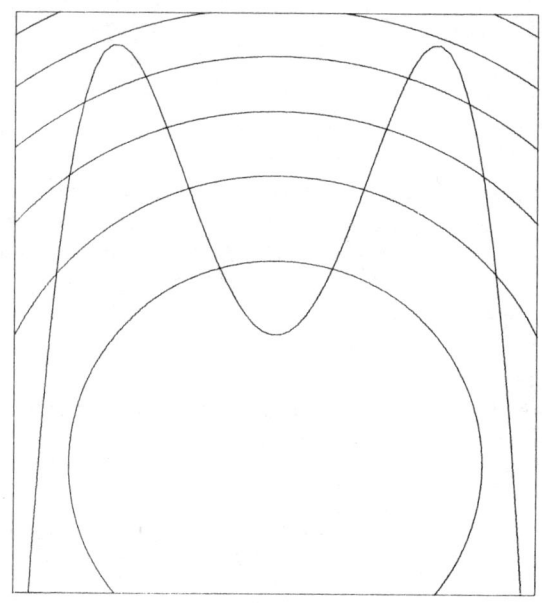

Fig. 1.8 Contours of the objective, and feasible set for P3

We next consider the functions

$$P_r(x) = (x^2 + y^2) + \frac{1}{r}(y + x^2(x^2-3)-1)^2 \qquad (1.75)$$

for $r = 10.0, 1.0, 0.1$ and 0.01. The graphs of these functions appear in Figs 1.9 to 1.12. One can observe that as r approaches zero, the penalty rises to a point that makes deep, curved valleys appear along the constraints, inducing the above-mentioned phenomenon of ill-conditioning for the corresponding unconstrained problem. Some spurious local minima may also appear, unrelated to the solution of the original problem. Clearly, the problem of minimizing $P_r(x)$ becomes increasingly more difficult and numerically awkward as the technique proceeds and as the approximations x_r^* approach the solution x^*.

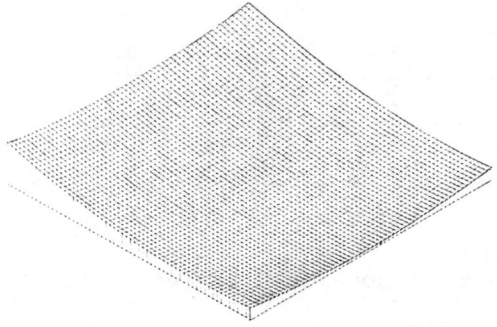

Fig. 1.9 Graph of penalized objective for P3: r = 100.0

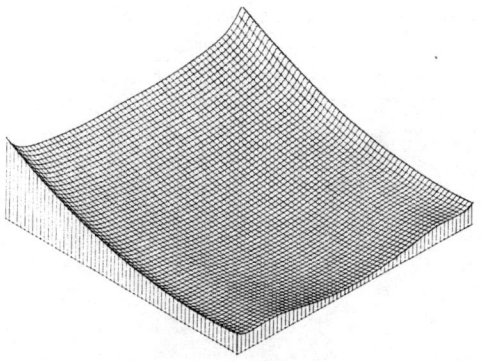

Fig. 1.10 Graph of penalized objective for P3: r = 10.0

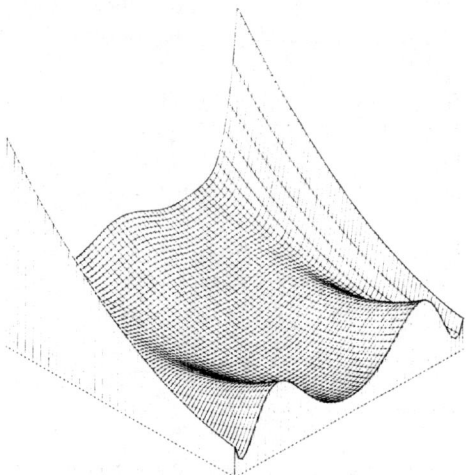

Fig. 1.11 Graph of penalized objective for P3: $r = 1.0$

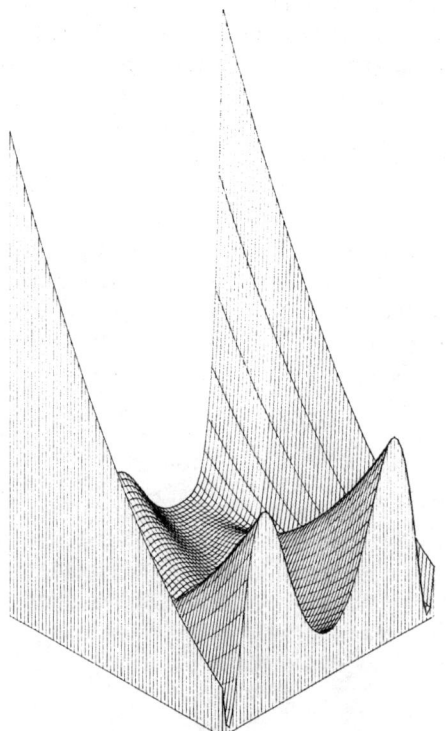

Fig. 1.12 Graph of penalized objective for P3: $r = 0.1$

Other penalty functions than (1.71) have been proposed, that use forcing functions other than the square of the constraint dissatisfaction. The exponential and absolute value function in particular have been considered. So called "interior penalty" or "barrier functions" have been designed for inequality constraints. The aim of these techniques is to keep feasibility by imposing on the objective function a strongly rising ridge inside the boundary of the feasible region. Although much ingenuity has been deployed in this area, none of the proposed techniques has escaped the curse of increasingly ill-conditioned subproblems. For this reason, the method is now less in use.

(b) Augmented Lagrangian methods

A way of avoiding the inherent numerical difficulties of penalty type methods is to design a sequence of functions whose unconstrained minima approaches the constrained solution without becoming too difficult to deal with. Suitable functions are a mixture of the Lagrangian (1.2) of the problem and of the penalty functions. Corresponding to (1.71), this class of functions, called augmented Lagrangians, are of the form

$$A(x,u,r) = f(x) - u^T e(x) + \sum_{i \in I_e} \frac{e_i(x)^2}{r_i} . \qquad (1.76)$$

(Inequalities may be dealt with by introducing slack variables, or by other methods we will not describe here.) Observe that (1.76) depends upon the knowledge of a certain u that represents the current Lagrange multipliers.

Despite the fact that this vector has to be computed somehow, (1.76) has over (1.71) the big advantage that r_i does not have to go to zero for the unconstrained minimum of (1.76) to coincide with the constrained solution. To illustrate this point, consider again problem (1.72)-(1.73). Another easy calculation shows that the optimal Lagrange parameter is given by

$$u = -2 \qquad (1.77)$$

Fig. 1.13 gives the contour lines (at the same heights as before) for (1.76) where r has been taken as infinite, i.e. when no penalty at all was applied and where the optimal value (1.77) was used.
One can immediately see that the solution (0,1) is the minimum

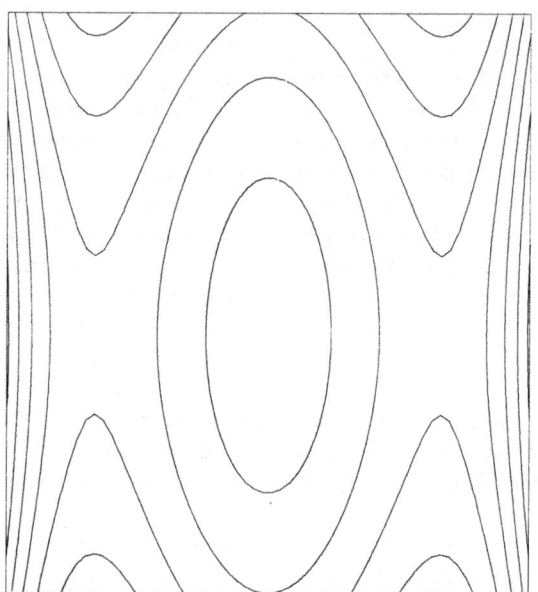

Fig. 1.13 Contours for the augmented Lagrangian for P3: r =

of $A(x,-2,+\infty)$. Consequently, there is no need to increase the penalty and worsen the (quite good) conditioning of the given function. In fact, it is possible to prove that we only need to decrease $\{r_i\}$ to a point where $A(x,u,r)$ is convex in a neighbourhood of the solution. Various schemes for computing the vector u have been proposed, involving a duality theory that will not be discussed here. Good surveys of these augmented Lagrangian or multiplier methods are given by Bertsekas [4] and by Rockafellar [30].

(c) Recursive quadratic programming methods

When opposed to primal methods, penalty or augmented Lagrangian methods show a rather undesirable feature: the dimension of the subproblems that are solved is always n and does not depend on the number of saturated or active constraints. An extreme example occurs when there are (n-1) equality constraints, so that there is just one degree of freedom left for the optimization problem. In this case, it seems preferable to solve a one-dimensional problem instead of a full n dimensional one.

Mathematical programming 29

A way to take this observation into account is to proceed by successive quadratic programs. That is, at the current iterate x, consider a problem with a quadratic objective function that approximates the behaviour of the Lagrangian in the x space, and the various constraints of the problems after they have been linearized. The saturation of the (linearized) constraints is then taken into account automatically by the quadratic program solver. It should also be noted that the solution of this quadratic program is a finite calculation, in contrast with the minimization that takes place in the above described methods.

It should be noted that the quadratic objective function approximates the Lagrangian (1.2) and not the objective function f(x) itself. This is because we are in fact solving equation (1.13). Indeed, these methods are related to Newton's algorithms for solving (1.13): the Lagrangian function incorporates some information about the constraint curvature that does not appear in the objective function. This is illustrated again using (1.72)-(1.73) where the actual iterate chosen is

$$x = 0.1 \quad , \quad y = 0.9 \quad , \quad u = -1.9 \; . \tag{1.78}$$

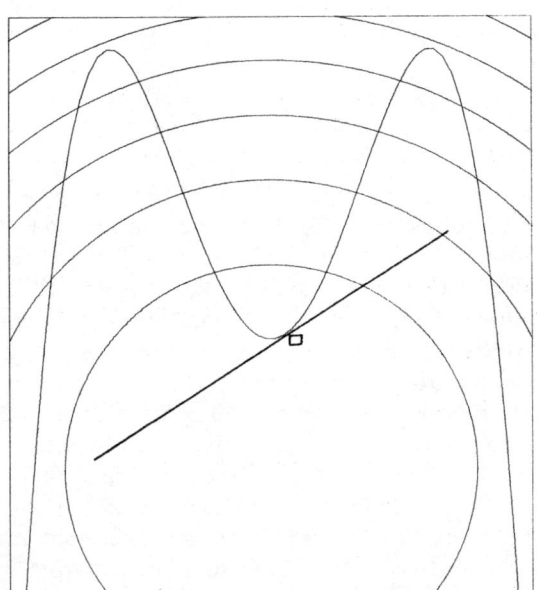

Fig 1.14 Contours of the objective for P3, with linearized constraints.

Fig 1.14 shows the contour lines of the objective function

together with the linearized constraint at the point (1.78). Fig. 1.15 shows the level lines of the Lagrangian (1.2) instead of the objective function (1.72). The point (1.78) is marked with a small square. Information about $\nabla^2 L(x,u,x)$ is usually obtained by using one of the known Quasi-Newton formulae that have been surveyed in 1.3.2(b). On the other hand, estimates for the Lagrange parameters are a natural by-product of the quadratic subprograms.

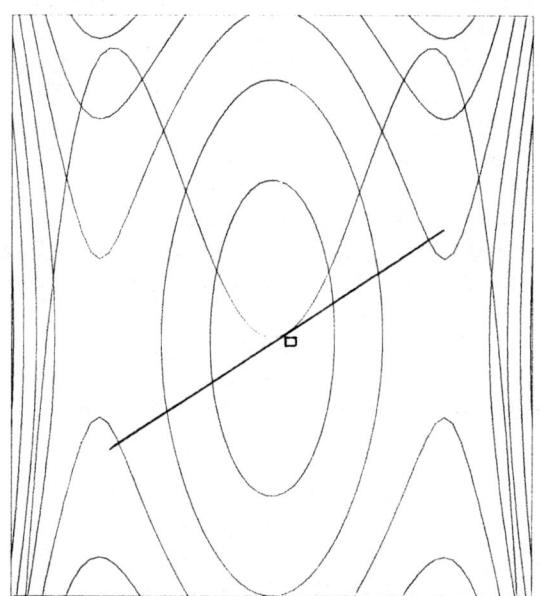

Fig. 1.15 Contours of the Lagrangian for P3.

Finally, there are two different ways of incorporating inequality constraints in the procedure. The first and conceptually simpler is to linearize them as the equality ones and to leave the quadratic program solver to sort out which ones are active. However, this has the drawback of making the quadratic program calculations a bit complicated. The second possibility is, for a given iterate, to decide a priori which constraints are saturated and to linearize these only. This helps to simplify the quadratic progamming routine but one has to design a so called "active set strategy" to make the decisions!

Work along these lines has been done by Biggs [5], Han [16], Powell [26], Coleman [8] and others. Their results show that this technique is very promising from the numerical point of view, although some design problems remain. (For example, what strategy has to be applied when the linearized constraints are incompatible, even if the true constraints are not?)

1.5 SOME OTHER MATHEMATICAL PROGRAMMING PROBLEMS

Sections 1.2-1.4 touch on only a restricted number of topics. We now discuss briefly some of the problems that were not mentioned in the previous sections and that are especially interesting.

1.5.1 Linear and Quadratic Problems

In this class of mathematical programs, the constraints are always linear and the function is either linear or quadratic. The most famous algorithm in this area is the simplex algorithm. It explores the vertices of the constrained set in an orderly manner to find the optimum. This procedure has been much refined in many directions, and a serious survey of linear programming is a very considerable task. An introduction to the area is given in chapter 4. Algorithms for solving quadratic problems are mostly derived from the linear programming field and also explore the feasible region in a suitable way. Procedures for both problems are highly numerically efficient.

1.5.2 Integer Programming

In some applications, it happens that some components of the vector x^* have to be integers. For example, they represent a number of items that cannot be split into parts (people, vehicles, etc.). Furthermore, it also happens that some variables can only assume the values 0 or 1, depending for example on the actual happening or not of some event. This introduces new constraints that are not of the type (1a)-(1b). Special techniques have been proposed for these problems, with varying numerical success. The only comment we make here is that this is a computationally very difficult problem that should be avoided if possible.

1.5.3 Non-Differentiable Programming

Another interesting class is that of problems having (locally) non-differentiable objective function or constraints. This may happen, for example, when some function is expressed as the maximum of several others. This topic has aroused a considerable theoretical and practical interest in

recent years. Most algorithms that have been proposed use
extended notions of differentiability that originated in the
study of convex functions (subdifferentiability, locally
Lipschitz functions, etc.). The theoretical aspect that has
been much developed, beside these notions, is optimality
conditions that generalize those presented in Section 1.2 to
the case of non-differentiable functions. From the practical
point of view, some algorithmic procedures have been proposed
for the unconstrained and more recently constrained case. We
nevertheless believe that those procedures should be used only
when classical differentiable routines fail. This remark is
based on the usually slow convergence rate of non-differentiable
methods.

ACKNOWLEDGEMENT

The author wishes to thank Dr. J. J. Strodiot for his comments
on an early draft of this chapter.

REFERENCES

[1] ABADIE J. and S. CARPENTIER (1966), Generalization of the Wolfe Reduced Gradient Method to the Case of Nonlinear Constraints in Optimization, R. Fletcher (Ed.), Academic Press.

[2] ALLWRIGHT J.C. (1972), In "Recent Mathematical Developments in Control", Bell D.J. (Ed.), Academic Press.

[3] BEALE E.M.L. (1972), In "Numerical Methods for Nonlinear Optimization", Lootsma F. A. (Ed.), Academic Press.

[4] BERTSEKAS D.P. (1976), Multiplier Methods, A Survey, Automatica, Vol. 12, pp. 133-145.

[5] BIGGS M.C. (1972), In "Numerical Methods for Non Linear Optimization", Lootsma F. A. (Ed), Academic Press.

[6] BRENT R.P. (1973), Algorithms for Minimization Without Derivatives, Prentice Hall, Englewood Cliffs, New Jersey.

[7] BUCKLEY A.G. (1978), A Combined Conjugate-Gradient Quasi-Newton Minimization Algorithm, Mathematical Programming, Vo. 15, pp. 200-210.

[8] COLEMAN T.F. (1979), Ph.D. Thesis, University of Waterloo, Ontario.

[9] DENNIS J.E. and MORE J.J. (1977), Quasi-Newton methods, Motivation and Theory, SIAM Review, Vol. 19, pp. 46-89.

[10] DIXON L.C.W. (1975), Conjugate Gradient Algorithms: Quadratic Termination Without Line Searches, J. of Inst. of Maths. and Appl., Vol.15, pp. 9-18.

[[11] FARKAS J. (1901), Über die Theori der Eingachen Ungleichungen, J. Reine Angew. Math., Vol. 124, pp. 167-205.

[12] FLETCHER R. and REEVES C.M. (1964), Function Minimization by Conjugate Gradients, Computer Journal, Vol. 7, pp. 149-154.

[13] FIACCO A.V. and McCORMICK G.P. (1968), Nonlinear Programming: Sequential Unconstrained Minimization Techniques, J. Wiley and Sons.

[14] GILL P.E. and MURRAY W. (1979), Private Communication.

[15] GOLDFARB D. (1969), Extension of Davidon's Variable Metric Method to Maximization Under Linear Inequality and Equality Constraints, SIAM J. Appl. Math., Vol. 17 pp. 739-764.

[16] HAN S.P. (1967), A Globally Convergent Method for Nonlinear Programming, J.O.T.A., Vol. 15, pp. 319-342.

[17] HUANG H.Y. (1970), Unified Approach to Quadratically Convergent Algorithms for Function Minimization, J.O.T.A., Vol. 5, No. 6, pp. 405-423.

[18] MANGASARIAN O.L. (1979), Nonlinear Programming, McGraw-Hill Book Co.

[19] MANGASARIAN O.L. and FROMOWITZ S. (1967), The Fritz John Necessary Optimality Conditions in the Presence of Equality and Inequality Constraints, J. Math. Anal. Appl., Vol. 17, pp. 37-47.

[20] MARWILL E. (1978), Ph.D. Thesis, Cornell University, Ithaca, New York.

[21] NAZARETH L. (1977), Conjugate direction algorithms without line searches, J.O.T.A., Vol. 23, No. 3, pp. 373-387.

[23] POLAK E. and RIBIERE G. (1969), Note sur la Convergence de Méthodes de Directions Conjuguées, Rev. Fr. Inform. Rech. Operation., Vol. 16-R1, pp. 35-43.

[24] POWELL M.J.D. (1964), An Efficient Method for Finding the Minimum of a Function of Several Variables Without Calculating Derivatives, Comp. Journal, Vol. 17, pp. 155-162.

[25] POWELL M.J.D. (1975), Restart Procedures for the Conjugate Gradient Method, Harwell Report CSS24.

[26] POWELL M.J.D. (1978), A Fast Algorithm for Nonlinearly Constrained Optimization Calculations, in Proceedings of the Dundee 1977 Numerical Analysis Meeting, Springer Verlag, pp. 144-157.

[27] POWELL M.J.D. and TOINT Ph. L. (1979), On the Estimation of Sparse Hessian Matrices, SIAM J. on Numer. Anal., Vol. 16.

[28] POWELL M.J.D. (1979), Gradient Conditions and Lagrange Multipliers in Nonlinear Programming, Proceedings of the meeting Optimization Techniques and Applications, held at Bergamo, to appear (Springer Verlag).

[29] RITTER K. (1975), A Method of Conjugate Directions for Linearly Constrained Nonlinear Programming Problems, SIAM J. on Numer. Anal., Vol. 12, pp. 273-303.

[30] ROCKAFELLAR R.T. (1976), Solving a Nonlinear Programming Problem by Way of a Dual Problem, Symposia Mathematica, Vol. 27.

[31] SHANNO D.F. and PHUA K.H. (1978), Matrix Conditioning and Nonlinear Optimization, Mathematical Programming, Vol. 14, pp. 149-160.

[32] SHANNO D.F. (1978), On the Convergence of a New Conjugate Gradient Algorithm, SIAM J. on Numer. Anal., Vol. 15, No. 6 pp. 1247-1257.

[33] SHANNO D.F. (1978), On Variable Metric Methods for Sparse Hessians, MIS Report No. 26, University of Arizona.

[34] SHANNO D.F. and MARSTEN R.E. (1979), Conjugate Gradient Methods for Linearly Constrained Nonlinear Programming, MIS Report No. 79-13, University of Arizona.

[35] STEWART G.W. (1967), A Modification of Davidon's Minimization Method to Accept Difference Approximations of Derivatives, J. ACM, Vol. 14, pp. 72-83.

[36] TOINT Ph. L. (1977), On Sparse and Symmetric Matrix Updating Subject to a Linear Equation, Maths, of Comp., Vol. 31, No. 140, pp. 954-961.

[37] TOINT Ph. L. (1979), A Note on Sparsity Exploiting Quasi-Newton Methods, to appear in Mathematical Programming Study on Unconstrained Optimization.

[38] ZANGWILL W.I. (1967), Minimizing a Function Without Calculating Derivatives, Computer Journal, pp. 293-296.

Chapter 2

Dynamic optimization

C. Storey

2.1 INTRODUCTION

In dynamic optimization, the problem of the determination of extrema of functionals is considered, as opposed to the extrema of functions in static optimization. A functional is a mapping from some function space into the real numbers, and so to find its extreme value a whole function has to be selected. This obviously brings in a number of mathematical difficulties not encountered in static optimization. It is possible, of course, to consider the optimization of functionals as a limiting case of the optimization of functions, as the number of variables tends to infinity, but methods have also been developed for dealing with the functional problem in its own right. In the next two chapters a brief account will be given of the calculus of variations and optimal control theory. The treatment can only be quite brief in the space available, and the reader is referred to the texts cited in the references for much fuller treatments.

Although problems in the calculus of variations and optimal control are not really distinct, it is convenient to deal with them separately.

2.2 THE CALCULUS OF VARIATIONS

Typical problems in the calculus of variations are:

(i) Find the shortest curve y(x) connecting two points A, B in a plane (Fig 2.1).

(ii) Find the curve y(x) joining two points A and B, in the first quadrant of a plane, such that the surface area obtained by rotating y(x) about the x-axis is minimized (Fig 2.2).

(iii) Find the curve y(x) joining two points A and B along which

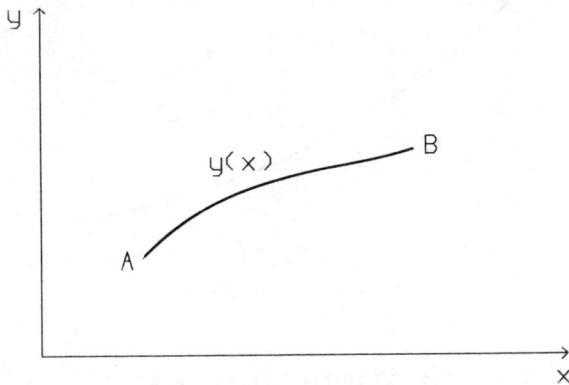

Fig 2.1   Shortest curve joining two points.

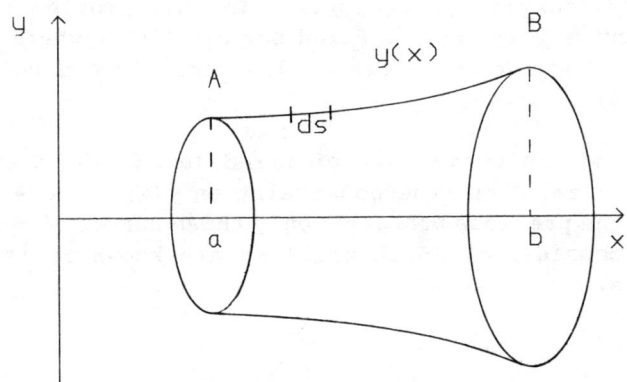

Fig 2.2   Surface of revolution of minimum area.

a particle would slide under gravity, without friction, from A to B in minimum time (Fig 2.3). This problem is known as the brachistochrone problem, and was first suggested by John Bernoulli in 1696. A more practical problem of this nature occurs in designing chutes so that the material passing through them has maximum exit velocity (Charlton et al, [9]).

(iv) The problem of streamlining. What shape must a body take to minimize its resistance when it is placed in a flow of fluid? Again a practical problem of this type which is of current interest involves the shape of bodies placed in the sea in order to extract the maximum amount of energy from the wave motion.

38 Dynamic optimization

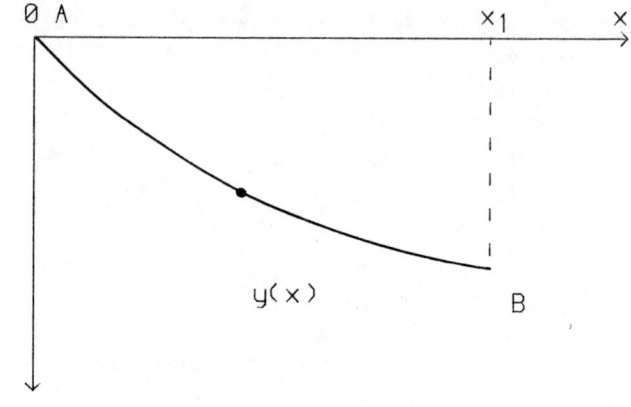

Fig 2.3   The brachistochrone problem.

(v)   Find that curve y(x) which minimizes the distance between two given curves $\phi_1(x)$, $\phi_2(x)$. In this problem the two ends of the curve y(x) are not fixed but can lie anywhere on the given curves.   Such problems are called variable end-point problems (Fig 2.4).

(vi) Find the curve y(x) of fixed length which encloses the maximum area. Here the constraint on y(x) is more restrictive than in the previous problem – only those curves of a given length can be considered.   Such problems are known as isoperimetric problems.

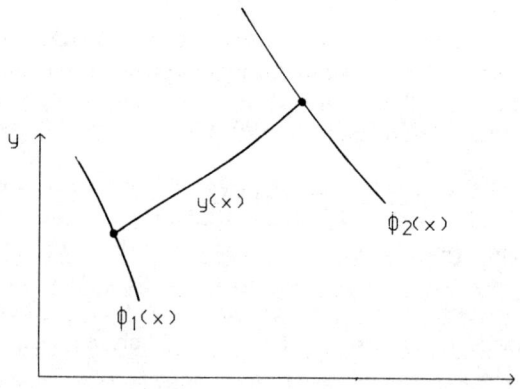

Fig 2.4   Variable end-point problem.

To illustrate the functional nature of the above problems, consider (ii). It is easy to see that if the area of the surface of revolution is denoted by I{y(x)}, to show its dependence on y(x), then

$$I\{y(x)\} = 2\pi \int_a^b y(x) \, ds$$

$$= 2\pi \int_a^b y(x) \sqrt{(1+y'(x)^2)} \, dx$$

where $y'(x)$ is dy/dx.

If, in particular, $a = 1$, $b = 2$, $y(1) = 1$, $y(2) = 2$ and $y(x) = x$, then

$$I\{y(x)\} = 2\pi\sqrt{2} \int_1^2 x \, dx \approx 13.329 \, .$$

On the other hand if, with the same end points,

$$y(x) = \tfrac{1}{3} x^2 + \tfrac{2}{3} ,$$

then

$$I\{y(x)\} = 2\pi \int_1^2 (\tfrac{1}{3} x^2 + \tfrac{2}{3}) \sqrt{(1 + \tfrac{4}{9} x^2)} \, dx$$

$$\approx 13.141 \, .$$

Other functions $y(x)$ would give different values of $I\{y(x)\}$, and problem (ii) is concerned with finding a $y(x)$ which makes the value of the functional $I\{y(x)\}$ a minimum.

2.2.1 The Euler Equations

As we have seen, many problems of the calculus of variations can be expressed in the following form. From some specified class of functions, find the function $y(x)$ which will minimize (or maximise) the integral

$$I\{y(x)\} = \int_a^b f(x, y(x), y'(x)) \, dx ,$$

where f is a given function of the three variables x, y and y'. The numbers a, b, $y(a)$ and $y(b)$ are assumed to be given. This is known as the simplest problem of the calculus of variations.

The class of functions from which the minimizing function can be selected differs from problem to problem and is called the class of admissible functions. For a given problem, it may be that no minimizing curve exists in some particular class of functions, but such a curve will exist if the class of admissible functions is suitably extended. For example, it is clear that within the class of continuous functions, the integral

$$\int_{x_0}^{x_1} y^2 dx$$

cannot have a minimizing curve unless $y(x_0) = y(x_1) = 0$.

The well-known necessary condition for a function $f: \mathbb{R}^n \to \mathbb{R}$ to have a maximum or a minimum at x^* is that $\nabla f|_{x^*} = 0$. By similar, though more complicated reasoning (for a rigorous account see for example, Sagan [36]), it can be shown that the analogous condition for the functional $I\{y(x)\}$ is that its first variation should vanish. Taking as the class of admissible functions C^2, the class of functions which are twice continuously differentiable, the first variation of $I\{y(x)\}$ is defined as

$$\frac{d}{d\varepsilon} I\{y(x) + \varepsilon \eta(x)\}|_{\varepsilon=0},$$

where $\eta(x)$ is a twice continuously differentiable function, with $\eta(a) = \eta(b) = 0$, but otherwise arbitrary.

Since

$$I\{y + \varepsilon \eta\} = \int_a^b f(x, y + \varepsilon \eta, y' + \varepsilon \eta') dx,$$

it follows that

$$\frac{d}{d\varepsilon} I\{y + \varepsilon \eta\} = \int_a^b \{\frac{\partial f}{\partial y} \eta + \frac{\partial f}{\partial y'} \eta'\} dx$$

where $\phi = y + \varepsilon \eta$. Hence the vanishing of the first variation implies

$$\int_a^b \{\frac{\partial f}{\partial y} \eta + \frac{\partial f}{\partial y'} \eta'\} dx = 0.$$

Integrating the second term by parts and using $\eta(a) = \eta(b) = 0$ gives

$$\int_a^b \{\frac{\partial f}{\partial y} - \frac{d}{dx}(\frac{\partial f}{\partial y'})\} \eta dx = 0.$$

It can be seen intuitively (and demonstrated rigorously) that since η is arbitrary, this implies

$$\frac{\partial f}{\partial y} - \frac{d}{dx}(\frac{\partial f}{\partial y'}) = 0. \tag{2.1}$$

A slightly different analysis can be carried out when the admissible functions have only piece-wise continuous derivatives, to give the necessary condition

$$\frac{\partial f(x,y(x),y'(x))}{\partial y'} = \int_a^x \frac{\partial f(t,y(t),y'(t))}{\partial y} dt + c, \quad (2.2)$$

(where c is a constant) for all $x \in [a,b]$ except the points of discontinuity of $y'(x)$. Clearly (2.1) holds for any smooth part of $y(x)$ and (2.1) and (2.2) are equivalent when $y'(x)$ is continuous. Equation (2.1) is known as the Euler equation for the simplest problem of the calculus of variations and (2.2) is known as the Euler equation in integrated form.

On carrying out the differentiation with respect to x the Euler equation (2.1) becomes

$$\frac{\partial^2 f}{\partial y'^2} \frac{d^2 y}{dx^2} + \frac{\partial^2 f}{\partial y \partial y'} \frac{dy}{dx} + \frac{\partial^2 f}{\partial y' \partial x} - \frac{\partial f}{\partial y} = 0, \quad (2.3)$$

which is a second-order ordinary differential equation for the required function $y(x)$. Solutions for (2.3) which satisfy the given end conditions $y(a) = y_a$, $y(b) = y_b$ are called extremals, and since (2.3) is a necessary condition only extremals need be examined, to see whether they provide maxima or minima or neither. Sufficient conditions can also be found (see, for example, Sagan, [36]) but are quite difficult to apply, and often a direct appeal to the nature of the problem can help to resolve the nature of the extremals.

There are a number of special cases for which either closed-form solution of the Euler equation is possible, or it is possible to find a first integral. Consider, for example, the situation when the integrand f is independent of x. Equation (2.3) then becomes

$$f_y - y' f_{y'y} - y'' f_{y'y'} = 0, \quad (2.4)$$

where f_y means $\partial f/\partial y$ and $f_{y'}$ means $\partial f/\partial y'$. On multiplying throughout by y' equation (2.4) reduces to

$$\frac{d}{dx}(f - y' f_{y'}) = 0,$$

so that a first integral is

$$f - y' f_{y'} = c, \quad (2.5)$$

with c an arbitrary constant.

It is easy to see that the brachistochrone problem is of this special type. If y is measured vertically downward (Fig 1.3) the velocity v of the particle is given by

$$v = \sqrt{(2gy)},$$

that is,

$$\frac{ds}{dt} = \sqrt{(2gy)}$$

and

$$T = \frac{1}{\sqrt{(2g)}} \int_0^{x_1} \frac{\sqrt{(1+(y')^2)}}{\sqrt{y}} \, dx.$$

Here

$$f = \frac{\sqrt{(1+(y')^2)}}{\sqrt{y}},$$

and the first integral (2.5) becomes

$$\frac{\sqrt{(1+(y')^2)}}{\sqrt{y}} - \frac{(y')^2}{\sqrt{(y(1+(y')^2))}} = k,$$

for some constant k. This reduces to

$$\frac{dy}{dx} = \sqrt{\left(\frac{c-y}{y}\right)},$$

where $c = \frac{1}{k^2}$. Making the substitution $y = c \sin^2 \frac{t}{2}$ gives

$$x = \frac{c}{2}(t - \sin t)$$

$$y = \frac{c}{2}(1 - \cos t).$$

These are the parametric equations of a cycloid, which is the well-known solution of the brachistochrone problem.

The solid-of-revolution problem is clearly also of this type, and a similar analysis shows that the solution is a catenary in that case. It is possible to show that for the brachistochrone problem there is a unique extremal (cycloid) through any two given points and this gives an absolute minimum time of descent. Things are rather more difficult in the case of solids of revolution, since according to the relative positions of A and B there may be one, two or no extremals (catenaries) passing through them. These remarks illustrate the great care which must be taken with regard to sufficiency. For more details the reader is referred to the texts on the calculus of variations in the references.

2.2.2 Extensions to the simplest problem

So far, only integrals which depend on a scalar $y(x)$ and its derivative $y'(x)$ have been considered. When the integrand depends on more than one function, and higher derivatives of these functions than the first, similar necessary conditions to the Euler equation (2.1) can be found. An example will suffice to illustrate these more general conditions. The Euler equations for the functional

$$I\{y_1(x), y_2(x)\} = \int_a^b f(x, y_1(x), y_1'(x), y_2(x), y_2'(x), y_2''(x)) dx,$$

with $y_1(a)$, $y_1(b)$, $y_2(a)$, $y_2(b)$, $y_2'(x)$, $y_2''(b)$ given, are

$$\frac{\partial f}{\partial y_1} - \frac{d}{dx}\left(\frac{\partial f}{\partial y_1'}\right) = 0$$

$$\frac{\partial f}{\partial y_2} - \frac{d}{dx}\left(\frac{\partial f}{\partial y_1'}\right) + \frac{d^2}{dx^2}\left(\frac{\partial f}{\partial y_2''}\right) = 0 .$$

In a similar way, the necessary conditions for problems in which the integrand contains functions of more than one variable, and their partial derivatives, involve partial differential equations. Again an example will illustrate this idea. A necessary condition for stationarity of the integral

$$I\{u(x,y)\} = \iint_S f(x, y, u, \frac{\partial u}{\partial x}, \frac{\partial u}{\partial y}) \, dx \, dy$$

with $u(x,y)$ prescribed on the boundary of the closed region S is

$$\frac{\partial f}{\partial u} - \frac{\partial}{\partial x}(f_{u_x}) - \frac{\partial}{\partial y}(f_{u_y}) = 0 , \qquad (2.6)$$

where

44   Dynamic optimization

$$f_{u_x} = \frac{\partial f}{\partial(\partial u/\partial x)},$$

etc. This is a second order partial differential equation and solutions which satisfy the prescribed boundary conditions are sought.

An important special case of the necessary condition (2.6) occurs when

$$I\{u(x,y)\} = \iint_S \{(\frac{\partial u}{\partial x})^2 + (\frac{\partial u}{\partial y})^2 + 2uF(x,y)\}\,dx\,dy, \quad (2.7)$$

with $F(x,y)$ a given function. The condition then reduces to the Poisson equation

$$\frac{\partial^2 u}{\partial x^2} + \frac{\partial^2 u}{\partial y^2} = F(x,y). \qquad (2.8)$$

Notice that the minimization of the integral (2.7) can be regarded an alternative to the solution of the partial differential equation (2.8).

The treatment of further complications in the calculus of variations, such as variable end points, isoperimetric problems, etc., can be found in the references. An isoperimetric problem of a practical nature, which involves the optimal design of beam-columns (members of a structure acted on by axial compression and transverse load simultaneously), is discussed in Burley, [26].

2.3 OPTIMAL CONTROL

Typical problems in optimal control are:

(i) Steer a rocket from the earth to the moon in such a way that the minimum amount of fuel is used and the rocket makes a soft landing.

(ii) Find the temperature profile along a tubular chemical reactor to maximize the yield of product.

(iii) Minimize the time of interception of an aircraft and a missile.

(iv) Maximize the transfer across the interface between two streams of fluid flowing in opposite directions, by controlling their speeds of flow.

These four problems are characterised by optimization of some measure of performance of a system described by a set of differential equations (ordinary differential equations for (i)-(iii) and partial differential equations for (iv)).

2.3.1 The Minimum Principle

Consider then a system described by the vector differential equation

$$\dot{x}(t) = f(x(t), u(t)) ,\qquad (2.9)$$

with $u(t) \in \mathbb{R}^m$, the control vector. Let the components of f on the right-hand side of (2.9) be f_i, $i = 1,2,\ldots,n$. The initial conditions are given by $x(t_0) = x^{(0)}$ and the interval of integration is $t_0 \leq t \leq t_f$ with t_0 and t_f specified. In most practical problems the control vector is not allowed unlimited variation, but must lie in some subset Ω of $\mathbb{R}^m$. It is required to select a control from some admissible set (e.g. the set of piecewise continuous functions) so as to minimize the performance index

$$J\{u(t)\} = \Phi(x(t_f)) + \int_{t_o}^{t_f} F(x(t), u(t))dt ,\qquad (2.10)$$

where Φ and F are given continuous functions with continuous first partial derivatives.

If the control vector u(t) is not constrained it is posssible to proceed in a similar manner to that of the calculus of variations by regarding the system equations as constraints and using Lagrange multiplier functions as follows. Let

$$L = J + \sum_{i=1}^{n} \int_{t_o}^{t_f} \lambda_i(t)(f_i - \dot{x}_i)dt ,\qquad (2.11)$$

where $\lambda(t) = (\lambda_1(t), \lambda_2(t), \ldots, \lambda_n(t))^T$ is a vector of Lagrange multipliers. Assuming the extension to the functional case of theorem 2 of chapter 1, a necessary condition for optimality is then the vanishing of the first variation of L corresponding to admissible variations in u(t). This can be shown to lead to the following set of conditions:

$$x_i(t_0) = x_i^{(0)}, \qquad \dot{x}_i = \frac{\partial H}{\partial \lambda_i} , \qquad i = 1,2,\ldots,n \qquad (2.12)$$

46 Dynamic optimization

$$\lambda_i(t_f) = \{\frac{\partial \Phi}{\partial x_i}\}_{t=t_f}, \quad \dot{\lambda}_i = -\frac{\partial H}{\partial x_i}, \quad i=1,2,\ldots,n \quad (2.13)$$

$$\frac{\partial H}{\partial u_i} = 0, \quad t_0 \leq t \leq t_f, \quad i=1,2,\ldots,m, \quad (2.14)$$

where the function H is defined by

$$H = F(x,u) + \sum_{i=1}^{n} \lambda_i f_i, \quad (2.15)$$

and is called the Hamiltonian function for the problem. (For a rigorous derivation of (2.12)-(2.15), see for example Luenberger [30].)

Notice that in addition to the n state equations (2.12) there are now also n equations (2.13) for the Lagrange multiplier functions $\lambda_i(t)$. These latter are known as the adjoint (or co-state) equations and the λ_i as the adjoint (co-state) variables. Since the boundary conditions for the λ_i are given in t_f the whole problem is of mixed boundary type and complicated by the fact that the optimal control vector u(t) is obtained by satisfying (2.15) at each stage of the solution. A very simple example will serve as illustration.

Example. A system satisfies the scalar differential equation

$$\dot{x} = -x + u, \quad 0 \leq t \leq 1,$$

and x(0) = 0. Select the control u so that the performance index

$$J = \frac{1}{2} \int_0^1 (x^2 + u^2) \, dt$$

is minimized. This is a scalar, linear system with a quadratic performance index. The final time t_f is fixed, the control is unconstrained and $\Phi \equiv 0$.

From (2.15) the Hamiltonian H is given by

$$H = \frac{1}{2} x^2 + \frac{1}{2} u^2 + \lambda(-x + u)$$

so that from (2.13)

$$\dot{\lambda} = -x + \lambda, \quad \lambda(1) = 0.$$

Setting $H_u = 0$, as in (2.14), gives the optimal control $u^* = -\lambda$.

Putting this into the state equation and solving this and the adjoint equation simultaneously yields the analytical solution:

$$x(t) = c_1 e^{t\sqrt{2}} + c_2 e^{-t\sqrt{2}}$$

$$\lambda(t) = -c_1(\sqrt{2}+1)e^{t\sqrt{2}} + c_2(\sqrt{2}-1)e^{-\sqrt{2}t}$$

with the constants c_1 and c_2 given by

$$c_1 = \frac{(\sqrt{2}-1)e^{-\sqrt{2}}}{(\sqrt{2}+1)e^{\sqrt{2}} + (\sqrt{2}-1)e^{-\sqrt{2}}}$$

$$c_2 = \frac{(\sqrt{2}+1)e^{\sqrt{2}}}{(\sqrt{2}+1)e^{\sqrt{2}} + (\sqrt{2}-1)e^{-\sqrt{2}}}.$$

Thus the whole problem can be solved analytically in this simple case.

When the control $u(t)$ is constrained to lie in some subset Ω of $\mathbb{R}^m$ the necessary condition still requires the solution of the state and adjoint equations, with their respective boundary conditions, but now the optimal control vector must at each time minimize the Hamiltonian with respect to all controls lying in Ω. In this form the solution is known as the minimum principle (originally maximum principle because of slightly different definition of H) of Pontryagin.

2.3.2 Dynamic Programming

Another approach to the solution of the optimal control problem is known as dynamic programming and is based on the "principle of optimality", which states that: "An optimal policy has the property that, whatever the initial decision is, the remaining decisions must form an optimal policy with respect to the state resulting from the initial decision." The optimal value function $S(x,t)$ for the problem is defined as the value of the performance index J obtained by using the optimal control from the initial state $x(t) = x$ over the time interval $[t,t_f]$. By using the principle of optimality it can be shown that S must satisfy the equation

$$\frac{\partial S}{\partial t} + \min_{u \in \Omega} \{F(x,u) + \sum_{i=1}^{n} \frac{\partial S}{\partial x_i} f_i(x,u)\} = 0, \qquad (2.16)$$

48 *Dynamic optimization*

with boundary condition (from the definition of S)

$$S(x,t_f) = \Phi(x(t_f)) . \qquad (2.17)$$

Equation (2.16), which is known as the Hamilton-Jacobi equation, is a partial differential equation for $S(x,t)$ where at each instant u must be chosen to minimize the expression in brackets over all $u \in \Omega$. The effect of this minimization is to make u a function of x and t, and the resulting partial differential equation can then be solved (in principle) for S.

There is a close connection between these two approaches to the optimal control problem. Indeed if $S(x,t)$ is assumed to have second partial derivatives and we define

$$p_i = \frac{\partial S}{\partial x_i} ,$$

then the Pontryagin principle, with $\lambda_i = p_i$, can be shown to be equivalent to the characteristic equations of the partial differential equation (2.16). Of course, the minimum principle was originally obtained in a rigorous direct manner under wider conditions than imposed above. For proofs of the above results and generalizations of the optimal control problem as stated here the reader is referred to the selection of texts on optimal control given in the references.

In general, it is not possible to solve optimal control problems analytically, and their numerical solution will be the subject of the next chapter. In one special case, which is of considerable interest, an analytical solution can be obtained. Consider the problem of controlling a general, time varying, linear system so as to minimize a quadratic performance index (the so-called linear-quadratic (LQ) problem):

$$\text{LQ: minimize } J = \frac{1}{2} x^T(t_f) P x(t_f) + \frac{1}{2} \int_{t_o}^{t_f} \{x^T Q(t) x + u^T R(t) u\} dt$$

$$\text{subject to } \dot{x} = A(t)x + B(t)u ,$$

$$x(t_0) = x^{(0)} .$$

The elements of the matrices A, B are continuous functions of time; P and R(t) are real, positive definite, symmetric matrices and Q(t) is a real, positive semi-definite matrix. For this problem, the Hamiltonian is given by

$$H = \frac{1}{2} x^T Q x + \frac{1}{2} u^T R u + \lambda^T (Ax + Bu)$$

and so the optimal control is

$$u^* = -R^{-1} B^T \lambda \,. \qquad (2.18)$$

The state and adjoint equations can therefore be written

$$\dot{x} = Ax - BR^{-1} B^T \lambda \,, \quad x(t_0) = x^{(0)} \qquad (2.19)$$

$$\dot{\lambda} = -Qx - A^T \lambda \,, \quad \lambda(t_f) = Px(t_f) \,. \qquad (2.20)$$

Suppose now that

$$\lambda(t) = \Lambda(t) x(t) \,, \qquad (2.21)$$

where $\Lambda(t)$ is a matrix to be determined (that $\lambda(t)$ is in fact of this form can easily be demonstrated: see, for example, Barnett, [3]). The optimal control (2.18) must therefore take the form

$$u^* = -R^{-1} B^T \Lambda x \,, \qquad (2.22)$$

which is said to be of linear feedback form. Differentiating (2.21) gives

$$\dot{\Lambda} x + \Lambda \dot{x} - \dot{\lambda} = 0 \,,$$

which on substituting for $\dot{x}$ and $\dot{\lambda}$ becomes

$$(\dot{\Lambda} + \Lambda A - \Lambda B R^{-1} B^T \Lambda + Q + A^T \Lambda) x(t) = 0 \,. \qquad (2.23)$$

Hence the matrix Λ is determined from the matrix differential equation

$$\dot{\Lambda} + \Lambda A - \Lambda B R^{-1} B^T \Lambda + Q + A^T \Lambda = 0 \,, \qquad (2.24)$$

with boundary condition

$$\Lambda(t_f) = P \,. \qquad (2.25)$$

Equation (2.24) is known as a matrix Riccati differential equation and is a set of $n(n+1)/2$ (Λ is symmetric) first-order quadratic differential equations for the distinct elements of Λ.

In the special case when A, B, Q and R are time-independent and $t_f = \infty$ it can be shown that the optimal control is

$$u^* = -R^{-1}B^T \Lambda x,$$

where Λ now satisfies the algebraic equation

$$\Lambda B R^{-1} B^T \Lambda - A^T \Lambda - \Lambda A - Q = 0.$$

This last matrix equation is a set of $n(n+1)/2$ simultaneous quadratic equations for the distinct elements of Λ.

2.4 COMPUTATIONAL METHODS

Although the analytical techniques discussed in the previous sections are formal solutions of the problems posed, it is seldom possible to obtain a general solution in closed form and recourse must be had to numerical methods. Even the "analytical solution" of the linear problem with quadratic performance index involves the solution of the appropriate matrix Riccati equation, which has to be dealt with numerically except in a few very special cases where a full closed solution is possible. It is the aim of this chapter to give a brief description of some of the more important numerical techniques involved in the solution of optimal control problems.

2.4.1 Reduction to parametric form

There are a number of methods for reducing problems involving optimization of functionals to problems involving optimization of functions, to which the methods of chapter 1 can then be applied. Perhaps the most natural way is the method of "finite differences". For example, in the simplest problem of the calculus of variations, where we recall that it is required to minimize

$$I\{y(x)\} = \int_a^b f(x, y(x), y'(x)) dx,$$

this method would proceed as follows. The interval of integration $[a,b]$ is divided up into $n+1$ equal subintervals, of length h, with end points

$$a, x_1, \ldots, x_n, b$$

and $y(x)$ is replaced by the line joining the vertices

$(a,y(a))$, $(x_1,y_1),\ldots,(x_n,y_n)$, $(b,y(b))$

where $y_i = y(x_i)$, $i = 1,2,\ldots,n$. The integral $I\{y(x)\}$ is then approximated by the sum

$$\sum_{i=0}^{n} hf(x_i, y_i, \frac{y_{i+1} - y_i}{h}),$$

which is a function of the n variables $y_1, y_2, \ldots, y_n$, (taking $y(a)$, $y(b)$ as fixed).

It is interesting that many of the early workers in the calculus of variations used the limiting form as $n \to \infty$ of solutions obtained by the above method to establish the corresponding continuous results. As a practical computing technique, however, this form of direct discretization tends to be uneconomical in computing time (see for example Walder and Storey [39]).

A second approach is to assume a relatively simple expression containing a number of free parameters for the required function $y(x)$. The functional becomes a function of these parameters which are then chosen to minimize it. Examples are

$$y(x) = a_1 x + a_2$$

$$y(x) = a_1 e^{a_2 x}.$$

Of course, more than one expression may be used. Thus two straight lines intersecting in a variable point would involve four parameters as follows:

$$y(x) = a_1 x + a_2, \quad x \leq a_4$$

$$y(x) = a_3 + (a_1 - a_3) a_4 + a_2, \quad x > a_4.$$

It is important that any knowledge of the nature of the solution to a specific problem should be used in selecting the most appropriate approximations. In addition, care must be taken to satisfy any given boundary conditions of the problem. A full discussion of the use of this method on the optimal temperature profile problem can be found in Walder and Storey [39]. A finite difference approach to a quite general formulation of the optimal control problem is described in Sargent and Sullivan [37] and successful performance is reported on high dimensional problems arising in the optimal control of distillation columns.

A more systematic form of the above technique is known as the Ritz (or Rayleigh-Ritz) method (Burley [7]; Kantorovitch and Krylov [25]), and involves representing the function y(x) by a truncated series

$$y_n(x) = \sum_{i=1}^{n} c_i \phi_i(x) .$$

In the above equation the c_i are constant coefficients and the $\phi_i(x)$ are selected in such a way that the sum is an admissible function satisfying the given boundary conditions. It is possible to show that with suitable conditions, $y_n(x) \to y(x)$ as $n \to \infty$. Substitution of $y_n(x)$ into $I\{y(x)\}$ gives a function $I(c_1,\ldots,c_n)$ of the coefficients only. A necessary condition for stationarity of this function is then

$$\frac{\partial I}{\partial c_i} = 0, \quad i = 1, 2, \ldots, n.$$

An example will illustrate the process.

Example

$$\text{Minimize} \quad I\{y(x)\} = \int_1^2 x^2 (y'(x))^2 dx$$

with

$$y(2) = \frac{1}{2}, \quad y(1) = 1.$$

Take

$$y(x) = -\frac{1}{2}x + \frac{3}{2} + (x-1)(x-2)(c_1 + c_2 x + \ldots c_n x^{n-1}),$$

where the linear term satisfies the actual boundary conditions and the remaining terms satisfy the corresponding homogeneous boundary conditions. With $n=1$, the approximation for $y(x)$ becomes

$$y_1(x) = c_1 x^2 - (3c_1 + \frac{1}{2})x + 2c_1 + \frac{3}{2} .$$

Hence

$$y_1'(x) = 2c_1 x - (3c_1 + \frac{1}{2}),$$

and some straightforward algebra gives

$$I(c_1) = \frac{4}{5} c_1^2 - \frac{1}{2} c_1 + \frac{7}{12}.$$

Setting

$$\frac{\partial I(c_1)}{\partial c_1} = 0$$

gives

$$c_1 = \frac{5}{16},$$

so that

$$y_1(x) = \frac{5}{16} x^2 - \frac{23}{16} x + \frac{17}{18}.$$

For this simple example the Euler equation can be solved analytically to give the exact solution.

$$y(x) = \frac{1}{x}.$$

The goodness of fit between the approximate and exact solution is shown in Table 1.

Table 1

x	1.00	1.20	1.40	1.60	1.80	2.00
$y_1(x)$	1.00	0.85	0.73	0.63	0.55	0.50
$y(x)$	1.00	0.83	0.71	0.63	0.56	0.50

The exact value of the integral is 0.5000 and the approximate value is 0.5052. It is apparent that a good result has been obtained for this example with only a single unknown parameter. (For fuller details of this problem see Pars [27].)

Of course, it is not always possible to find algebraic expressions for $\partial I/\partial c_i$ or to solve the equations $\partial I/\partial c_i = 0$, $i = 1, 2, \ldots, n$. It is always possible, however, in such difficult cases to resort to solving the problem completely by numerical methods, that is, by the use of numerical integration and optimization with respect to the c_i. The direct numerical approach may also give more scope for the selection of the trial functions $\phi_i(x)$ to suit the nature of the problem. For some examples of direct numerical solution of engineering problems see Rosenbrock and Storey [35] and Pakes and Storey [32]. For an elementary account of other direct methods in the calculus of variations, Burley [7] should be consulted.

A form of direct method which has been developed specifically for optimal control problems is known as the "ε-method of Balakrishnan". This method is analogous to the penalty function methods of chapter 1.4, with the state equations treated as constraints, and is organised on the following lines. Suppose the problem is to minimize

$$J = \Phi(x(t_f)) + \int_{t_0}^{t_f} F(x(t),u(t),t)dt$$

subject to the state equations

$$\dot{x}(t) = f(x(t),u(t),t),$$

with initial condition

$$x(t_0) = x_0$$

and time interval $t_0 < t < t_f$. Set up a new cost function

$$K = \Phi(x(t_f)) + \frac{1}{2\varepsilon} \int_{t_0}^{t_f} \|\dot{x}(t) - f(x,u,t)\|^2 dt + \int_{t_0}^{t_f} F(x,u,t)dt$$

and minimize this by selecting suitable functions x and u (as in the method of Ritz). The whole procedure is then repeated for a sequence of decreasing values of ε. Clearly the new cost function is constructed in such a way that approximate satisfaction of the state equations is enforced. Fuller details of Balakrishnan's method and its use can be found in Balakrishnan [2], Huang [24] and Connor [12].

2.4.2 Gradient methods

A major difficulty with dynamic optimization problems is that they generally give rise to two-point boundary value problems: the state variables are specified at the initial time t_0 and the adjoint variables are specified at the final time t_f. One approach to this problem is to iterate in some manner on the unknown values of, say, the adjoint vector at t_0 until a suitable error criterion involving its given values at t_f is met. Such boundary iteration methods, however, require that the two sets of equations (state and adjoint) should be integrated in the same direction, either forward from t_0 or backward from t_f. This gives rise to numerical instability in one set of equations or the other.

To overcome this difficulty methods have been developed in which both sets of equations are integrated in their naturally stable directions. One of the earliest of these methods is analogous to the method of steepest descents in mathematical programming. Consider the optimal control problem of of the previous chapter with $F \equiv 0$, and u taken to be scalar for simplicity. Given an approximation u_0 to the optimal control then a better approximation, in the sense of taking a step along the path of steepest descent of the functional $\Phi(x(t_f))$, can be shown to be

$$u_1 = u_0 + \sigma \frac{\partial H}{\partial u}$$

where H is the Hamiltonian and the parameter σ determines the length of step taken.

The computational procedure is now as follows. With an initial guess $u_0(t)$ at the control the state equations are integrated from t_0 to t_f and the adjoint equations are integrated from t_f to t_0. At each stage of the integration the nominal control u_0 is adjusted by the addition of $\sigma \partial H/\partial u$. The whole process is then repeated for the new control u_1. Notice that the instability inherent in the methods that use some form of iteration on the boundary values is avoided by this gradient method. An account of numerical experience concerning the selection of σ and the effect of different numerical integrating methods can be found in Walder and Storey [39]. For a survey of the use of the gradient method in more general control problems and its convergence properties see Rosenbrock and Storey [35], Bryson and Denham [5] and Connor [13]. Just as is the case for function optimization, the method of steepest descent is relatively inefficient but has the advantage of not needing an excessive amount of computation to obtain the step direction.

It is also possible to extend many of the other techniques of function optimization to functional problems. For example, the Fletcher-Reeves form of the conjugate gradient algorithm (chapter 1.3.2) extends in the following manner to the optimal control problem just discussed. Denote $\partial H/\partial u$ by $g(u)$. Then for the ith iteration,

(i) $u_{i+1} = u_i + \sigma_i s_i$, $i = 0,1,2,\ldots$ with the scalar σ_i chosen so that $J\{u_{i+1}\}$ is minimized. The initial control u_0 is arbitrary and $s_0 = -g(u_0)$.

(ii) $s_{i+1} = -g(u_{i+1}) + \beta_i s_i$, $i = 0,1,2...$

where

$$\beta_i = \frac{<g(u_{i+1}),g(u_{i+1})>}{<g(u_i),g(u_i)>}$$

and $<.,.>$ is the appropriate inner product.

To obtain g(u) at each iteration it is, of course, necessary to integrate the state and adjoint equations in a similar way to that in the method of steepest descent. The conjugate gradient method can also be applied to more complicated control problems than the simple one considered above. For numerical comparison Walder and Storey [39], Connor and Saltavarias [14] and Garg [20] should be consulted.

It is natural to attempt to improve the convergence of the techniques discussed above by the use of second order methods as in function optimization (Newton, variable metric, etc.). There is a brief account of a number of such methods with references to the original papers in Connor [13]. This paper also describes a number of other numerical algorithms including differential dynamic programming and the use of contraction mappings. In Turner and Huntley [38] some of the results concerning the variable metric methods used in function optimization are extended to functionals on Hilbert spaces and then used in a computational study of a control problem.

There is a detailed survey of functional minimization in Garg [20], together with numerical comparison of a new second order gradient technique with gradient and conjugate gradient first order methods, Davidon's variable metric method, quasi-linearization and two extensions of Davidon's rank-one method.

2.5 OTHER AREAS OF RESEARCH

Clearly it is possible to give only a very brief coverage of functional optimization in this chapter. Connor [13] may be consulted for a survey of work on hereditary systems (that is for which the state equation involves time delays), state space constraints, and the use of penalty functions. For singular controls and the numerical techniques associated with these problems see Bell and Jacobson [4].

There is also considerable research activity in optimal control of distributed systems (systems governed by partial differential equations), for example Lions [28], Davies [16];

and in the optimization problems arising in the study of stochastic systems: Fuller [18], Kushner [27]. A broad idea of some recent research in optimal control theory and computation (Chapters 3 and 4) and in stochastic control (Chapter 5) can be obtained from Gregson [22].

REFERENCES

[1] M. Athans and P.L. Falb. Optimal control, McGraw Hill, 1966.

[2] A. V. Balakrishnan. On a new computing technique in optimal control and its application to minimal-time flight profile optimization, J.O.T.A., $\underline{4}$, No.1, 1969.

[3] S. Barnett. Introduction to Mathematical Control Theory, Clarendon Press, Oxford, 1975.

[4] D. J. Bell and D. H. Jacobson. Singular optimal control problems, Academic Press, 1975.

[5] A. E. Bryson and W. F, Denham. A steepest ascent method for solving optimum programming problems, J. App. Mechs., p. 247, June 1962.

[6] A.E. Bryson and Y.C.Ho. Applied optimal control, Blaisdel, 1969.

[7] D.M. Burley. Undergraduate studies in optimization, International Textbooks, 1974.

[8] M.D. Canon, C.D. Cullum, E. Polak. Theory of optimal control and mathematical programming, McGraw Hill, 1970.

[9] W. Charlton, C. Charella, A.W. Roberts. Gravity flow of Granular materials in Chutes : Optimization flow properties, J. agric. Engng. Res. (1975), $\underline{20}$, 39-45.

[10] S.J. Citron. Elements of optimal control, Holt, Rinehart and Winston, 1969.

[11] J.C. Clegg. Calculus of variations, Oliver and Boyd, 1968.

[12] M. A. Connor. A maximum principle for neutral systems via Balakrishnan's ε-method, J.O.T.A., $\underline{12}$, No. 1, 1973.

[13] M. A. Connor. Some recent developments in optimal control theory, Bull. I.M.A., $\underline{12}$, Nos. 11/12, 1976, 347-354.

[14] M. A. Connor and P. Saltavarias. Augmented penalty

function methods in optimal control, to appear in Proc. IMA. Conference on Recent Theoretical Developments in Control, Leicester, Sept. 1976.

[15] J.W. Daniel. The approximate minimization of functionals, Prentice-Hall, 1971.

[16] T. V. Davies. A review of distributed parameter system theory, Bull. I.M.A., 12, No. 5, May 1976.

[17] S.E. Dreyfus. Dynamic programming in the calculus of variations, Academic Press, 1965.

[18] A.T. Fuller (Ed.), Non-linear stochastic control systems, Taylor and Francis, 1970.

[19] R.V. Gamkrelidze. Principles of optimal control theory, Plenum Press, 1978.

[20] S. C. Garg. Numerical minimization methods for functionals: Comparison and extensions, U.T.I.A.S. Report No. 209, July 1977.

[21] I.M. Gelfand and S.V. Fomin. Calculus of variations, Prentice-Hall, 1963.

[22] M. J. Gregson, (Ed.) Recent theoretical developments in control, Academic Press, 1978.

[23] M. R. Hestenes. Calculus of variations and optimal control theory, John Wiley, 1966.

[24] Sheng-Chao Huang. A constructive approach to the maximum principle for differential-difference problems using Balakrishnan's ε-technique] J.O.T.A., 5, No. 1, 1970.

[25] L. V. Kantorovich and V. I. Krylov. Approximate methods of higher analysis, Ch. 4, Interscience, 1958.

[26] B.L. Karihaloo and R.D. Parberry. The optimal design of beam-columns, Int. J. Solids Structures, Vol. 15, 1979, 855-859.

[27] H. Kushner. Introduction to stochastic control] Holt, Rinehart and Winston, 1971.

[28]   J. L. Lions.   Contrôle optimal de systèmes gouvernés par des équations aux dérivées partielles, Dunod, Paris, 1968.

[29]   E.B. Lee and L. Markus. Foundations of optimal control, John Wiley, 1967.

[30]   D.G. Luenberger. Optimization by vector space methods, John Wiley, 1969.

[31]   L.W. Neustadt.   Optimization: A theory of necessary conditions, Princeton University Press, 1976.

[32]   H. W. Pakes and C. Storey. Solution of the equations for a tubular reactor with axial diffusion by a variational technique, The Chemical Engineer, No. 208, 1967, CE96-CE108.

[33]   L.A. Pars. An introduction to the calculus of variations, Heinemann, 1962.

[34]   B.N. Pshenichnyi. Necessary conditions for an extremum, Marcel Dekker, 1971.

[35]   H. H. Rosenbrock and C. Storey. Computational techniques for chemical engineers, Pergamon Press, 1966.

[36]   H. Sagan.   Introduction to the calculus of variations, McGraw Hill, 1969.

[37]   R. W. H. Sargent and G. R. Sullivan. The development of an efficient optimal control package, in Optimization Techniques, Part 2, Lecture Notes in Control and Incormation Sciences, $\underline{7}$, Ed., J. Stoer, Springer-Verlag, 1978, 158-168.

[38]   P. R. Turner and E. Huntley. Variable metric methods in Hilbert space with applications to control problems, J.O.T.A., $\underline{19}$, No. 3, 1976, 381-399.

[39]   T. J. Walder amd C. Storey. Numerical solution of an optimal temperature problem, The Chemical Engineering Journal, $\underline{1}$, 120-128, 1970.

[40]   J. Warga. Optimal control of differential and functional equations, Academic Press, 1972.

[41]   L.C. Young. Lectures on the calculus of variations and optimal control theory, W.B. Saunders, 1969.

Chapter 3
Decomposition methods in optimization
G.W.T. White

3.1 INTRODUCTION

The effective management and control of large systems is difficult because, amongst other things, the task of formulating adequate models and objectives is very complex, and because sheer size makes the techniques for using the models and computing the optimal values of the objectives very slow or even impractical. Nevertheless the rewards for success are high, and during the last fifteen years considerable effort has been devoted to the analysis of large-scale problems and to the synthesis of effective techniques for solving these problems — see for example Himmelblau [11], or Proceedings of the IFAC Symposium on Large Scale Systems Theory and Applications, Udine, Italy (1976). As an area of research and application this challenging work has attracted the attention of the control engineer, the applied mathematician and the management scientist, and in the literature on large-scale systems we find contributions from many disciplines. A study of this literature reveals the underlying belief that the fundamental characteristic of a large, complex system is that it is not an amorphous aggregate but a purposefully interlinked assembly of units or subsystems. Likewise, it is the thesis of this chapter that systems are structured, and knowledge of this structure should be exploited in the modelling, control and management of the complex. Any global problem requiring solution for the whole complex should be broken down into a set of sub-problems, one associated with each subsystem; these subproblems are then solved independently of each other but under the influence of a coordinator whose task it is to account for the interconnections and conflicts between the subsystems. Thus we are concerned with the concepts of decomposition and coordination, of hierarchical control and of decentralization.

There is a lack of concensus about the precise meaning of the terms 'decentralized' and 'hierarchical'. In the literature, and to some extent here, the two terms are often used synonymously. Nevertheless, it is sometimes desirable to distinguish between the concept of a decentralized system and that of a hierarchical system, and there is growing support (Sandell, Varaiya and Athans [18]; Wilson [22]) for the following view. A centralized system is one in which all the system information is available centrally, and in which all the system variables may be manipulated directly from the centre. Conversely, in a decentralized system, the subsystems have available only strict subsets of the system information, and are able to manipulate only strict subsets of the system variables. A decentralized system is hierarchical only if the information subsets of some subsystems depend directly on the action of other subsystems, thus establishing a priority of intervention of some susbsystems over others. However, a hierarchical system is not decentralized if each subsystem may operate directly on all the system variables.

These concepts, and some of the problem solving techniques resulting from them, are applicable to many aspects of management science including information structures and decision making (Athans [1]; Ho and Chu [12]; Bailey [2]), scheduling (Drew [6]) and control (Findeisen [7]). In much of the work in these areas it is explicitly or implicitly assumed that the problems which arise can be posed as constrained, non-linear optimization problems such as the minimization of a cost functional or the maximization of some profit measure. Admittedly, in many practical cases the problem solutions may be sought by methods other than optimization, but it is believed that the information structures needed to solve the problems by decentralized hierarchical optimization will be similar to the structures that are needed for any other decentralized hierarchical solution method. Thus it is the purpose of this paper to assess the various decomposition and coordination techniques whereby optimization problems may be solved in a decentralized hierarchical fashion. The optimization problems can arise in many forms. The variables may be subject to stochastic constraints and the objective function may be an expectation, or the whole problem may be deterministic. The constraints may be differential equations and the objective function may be an integral with respect to time, i.e. the problem may be a dynamic one, or the problem may be static. However, the various strategies for decomposing an optimization problem do not necessarily depend upon the type of problem, and so for the purposes of introducing the basic decomposition and coordination techniques and delineating the associated structures of information flow, consideration of the deterministic static problems is sufficient and in this chapter attention is restricted to this type of problem; it is not only the simplest type of problem, and therefore the best suited to illustrate the methodology, but it is also the type of problem whose solution has the greatest potential for application.

3.2 THEORY OF DECENTRALIZED OPTIMIZATION

The aim of this section is to review sufficient of the basic theory of decentralized optimization to allow a discussion of the advantages and disadvantages of the technique and its associated algorithmic problems. The theory of decentralized optimization has been treated in depth in a number of authoritative works amongst which the books by Lasdon [15], Findeisen [7] and Singh and Titli [20] are notable. The first concentrates on large-scale linear and, to a lesser extent, non-linear problems arising typically in an operations research environment, and gives a detailed analysis of the decomposition algorithms of Dantzig and Wolfe [5], which is examined in chapter 5 of this book, Benders [3] and Rosen [17]. The other two are primarily concerned with separable optimization problems arising in multilevel control systems. The basic manipulations and strategies used in decomposing optimization problems have been reviewed by Geoffrion [10] and more recently by Wilson [22]. The theory is not new therefore, neither can a review as brief as that given below fully explain all the variations that are possible. What follows is a discussion of the most basic methods of decomposing and coordinating non-linear programming problems: the treatment follows closely that given by Simmons and White [19] and Findeisen [8].

To avoid difficulties of understanding arising from unnecessary mathematical abstraction, and to emphasise the relevance of the theory to real problems, we derive the mathematical programming problem which forms the starting point for our treatment of theory by considering a disaggregated production system.

3.2.1 A complex production system

Suppose an assembly of N subsystems (factories or process units) is interconnected by product streams, as shown schematically in Fig 3.1. The ith subsystem has a vector of inputs z_i from the other subsystems,

64 Decomposition methods

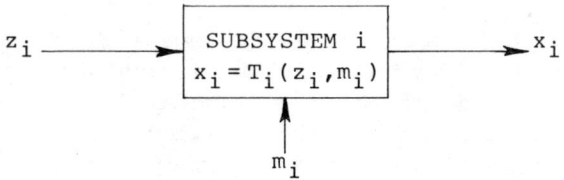

Fig 3.1 Subsystem i of an interconnected production system.

a vector of outputs x_i to the other subsystems and a vector of controls m_i. It is assumed that the algebraic equation describing the performance of the subsystem, that is the model of the subsystem, is

$$x_i = T_i(z_i, m_i), \qquad (3.1)$$

and that the feasible operating region of the subsystem is defined by the vector of local constraints

$$h_i(z_i, x_i, m_i) \leq 0. \qquad (3.2)$$

The interconnections between the subsystems are simple in the sense that an output stream can be characterised by a single variable and goes directly from one subsystem to another, although stream merging and recycling are allowed; under these conditions we may write

$$G \begin{matrix} z \\ x \end{matrix} = 0, \qquad (3.3)$$

where G is a matrix of ones and zeros, generally sparse and non square, and where z and x are the composite input and output vectors

$$z = (z_1^T, z_2^T, \ldots, z_N^T)^T$$

$$x = (x_1^T, x_2^T, \ldots, x_N^T)^T.$$

There are several ways in which the matrix G can be partitioned; for the methods described in this section it is convenient to proceed by suitably permuting the columns of G and ordering the elements of z and x to group together the inputs and outputs of each subsystem so that equation (3.3) may be written

$$[G_1 \quad G_2 \quad \cdots \quad G_N] \begin{bmatrix} z_1 \\ x_1 \\ z_2 \\ \cdot \\ \cdot \\ \cdot \\ x_N \end{bmatrix} = 0,$$

or

$$\sum_{i=1}^{N} G_i \begin{bmatrix} z_i \\ x_i \end{bmatrix} = 0 \tag{3.4}$$

For convenience of notation we write this as

$$\sum_{i=1}^{N} g_i(z_i, x_i) = 0, \tag{3.5}$$

where the g_i are linear vector functions having a dimension equal to the number of rows of G, i.e. equal to the total number of product streams. It is assumed that there is a scalar objective function for each subsystem, $f_i(z_i, x_i, m_i)$, and that the global objective function is just the sum of the subsystem objectives

$$\sum_{i=1}^{N} f_i(z_i, x_i, m_i). \tag{3.6}$$

Thus the global mathematical programming problem which we denote by GMP and which we assume to be well posed is

$$\text{GMP: Maximize}_{z,x,m} \sum_{i=1}^{N} f_i(z_i, x_i, m_i) \tag{3.6}$$

$$\text{subject to } \sum_{i=1}^{N} g_i(z_i, x_i) = 0$$

$$T_i(z_i, m_i) - x_i = 0, \quad i = 1, 2 \ldots, N$$

$$h_i(z_i, x_i, m_i) \leq 0, \quad i = 1, 2 \ldots, N$$

For later use let us define z^*, x^*, m^* to be the solution of this problem and let

$$\sum_{i=1}^{N} f_i(z_i^*, x_i^*, m_i^*) = F$$

What we seek now are ways in which GMP may be broken down into N independent optimizations which can be coordinated in some way to achieve the solution of the global problem. We describe below two basic methods: primal coordination and dual coordination. A study of these two methods is an essential basis for understanding the various decomposition and coordination procedures described in the literature.

3.2.2 Primal Coordination

In this method the solution of GMP is sought by maximizing the global objective function iteratively, first over the controls m, then over the inputs and outputs z and x, then again over m and so on until convergence is achieved. This is essentially Geoffrion's [10] method of projection, Mesarovic et. al's [16] method of model coordination and what Kulikowski et al. [14] and Wilson [22] call parametric decomposition. To analyse the conditions under which the method will work it is necessary to describe it more formally, which we may do as follows. If the interconnection variables z and x are fixed at some value which satisfies the interconnection constraint (3.5), then the maximization of (3.6) is only over the controls m, and the problem GMP separates into N independent maximization problems

$$\text{MP}(i): \underset{m_i}{\text{Maximize}} \quad f_i(z_i, x_i, m_i)$$

$$\text{subject to} \quad T_i(z_i, m_i) - x_i = 0 \quad (3.7)$$

$$h_i(z_i, x_i, m_i) \leq 0$$

Let the solutions of these problems be denoted by $\{m_i^+\}$ (more correctly by $\{m_i^+(z_i, x_i)\}$ because these solutions depend upon the parameters $\{z_i\}$ and $\{x_i\}$, but the inclusion of the arguments is notationally clumsy and is therefore omitted). The value of the global objective function for these values of $\{m_i\}$ (and the chosen values of $\{z_i\}$ and $\{x_i\}$) will be

$$\sum_{i=1}^{N} f_i(z_i, x_i, m_i^+) = \psi(z, x)$$

and the aim is to maximize this function. Thus the successive values of z and x will be determined by solution of the coordinator or master problem

CP: Maximize $\psi(z,x)$

$$\text{subject to} \quad \sum_{i=1}^{N} g_i(z_i, x_i) = 0 \tag{3.8}$$

$$(z,x) \in V$$

where the set V is introduced to ensure that it is possible to find solutions to the problems MP(i). The set V is defined as follows

$$V = \{(z,x): \text{there exist } m_i \text{ satisfying equations}$$

$$(3.7) \text{ for all } i\} \tag{3.9}$$

The problems MP(i) and CP are totally interdependent. In one iteration of the algorithm the local problems MP(i) take the values of z and x specified by CP and hand back the optimal value of the local objective functions $f_i(z_i, x_i, m_i^+)$; the coordinator uses these values to compute new feasible values of z, x which will increase ψ, and sends these values to the subsystems ready for the next iteration. A diagram of this information flow for a three subsystem problem is shown in Fig 3.2.

It should be noted that because CP is constrained to work with feasible z and x, i.e. values of z and x satisfying (3.5) the successive values of m_i^+ generated as the iterations progress will maintain the stream interconnections in balance, although not at their optimal values until the iterations have converged. This property, which is not possessed by the dual coordination method described below, is useful in on-line situations. Also in contrast to the later methods, primal coordination does not introduce any auxiliary variables to achieve decomposition and thus the sum of the dimensions (number of variables) of the problems CP and MP(i) is the same as the dimension of GMP. However, this very point is intrinsically an important source of difficulty with primal coordination - in working with the minimum number of variables the algorithm may not give

68   Decomposition methods

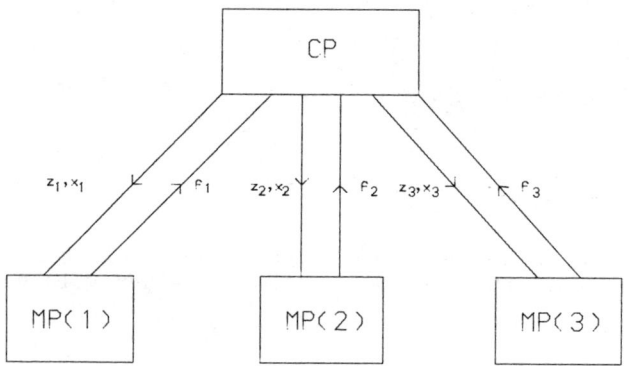

Fig 3.2   Information flow in primal coordination.

sufficient degrees of freedom to permit solution of the subsystem problems. It is quite possible for a subsystem to have more inputs and outputs than controls and thus if the inputs and outputs are arbitrarily fixed it is likely that there are no feasible values of m_i which can achieve satisfaction of the local constraints (3.7). This is part of the reason for introducing the set V into CP, but the simple definition of V given in (3.9) belies the fact that it is very difficult to compute this set at the coordinator level without centralizing all the local information (Findeisen [8]) although some resolution of this difficulty is possible if all the functions T_i and h_i are linear (Kulikowski et al. [14]).

Another source of difficulty is that, in general, the gradient of the objective function $\psi(z,x)$ is not easy to compute and may not even exist: this is an important limitation because one iteration of the coordinator problem requires solution of N problems {MP(i)}. Thus it is desirable to solve CP efficiently, and this is not really possible without gradient information. Even if assumptions are made that the functions f_i are strictly concave, the functions h_i are convex and differentiable and the functions T_i are linear, the function ψ may still not be differentiable, but it can be shown that ψ is concave. Therefore sub-differentials of ψ exist and may be used as a basis for effective optimization algorithms, although details of such algorithms for solving CP have been worked out only for rather special cases (Lasdon [15]).

These difficulties with primal coordination, which stem in part from the inflexibility of equality coupling constraints, constitute just the kind of difficulties which must be overcome in industry when coordination of a complex production system is attempted by specifying the product rates and feedstock consumptions of each factory or process unit. In practice these difficulties are largely overcome by the provision of product storage thus allowing some relaxation of the interconnection constraints and in Simmons and White [19] consideration is given to how this relaxation can be exploited in the context of decentralized optimization. Here an alternative method of coordination is presented.

3.2.3 Dual Coordination

Dual coordination is based on strong Lagrangian theory (see, for example, Whittle [21]). The coupling constraints (3.5) are adjoined to the objective function (3.6) using a vector of Lagrange multipliers of a dimension equal to the number of elements of g_i, i.e. equal to the number of interconnection streams. Thus a global Lagrangian function is defined as

$$L(z,x,m,\lambda) = \sum_{i=1}^{N} f_i(z_i,x_i,m_i) - \lambda^T \sum_{i=1}^{N} g_i(z_i,x_i) \qquad (3.10)$$

and the maximum of this function is sought subject only to the local constraints (3.1) and (3.2). This maximization problem we denote by ML.

ML: Maximize $\quad L(z,x,m,\lambda)$
$\quad\quad\;\; z,x,m$

$\quad$ subject to $T_i(z_i,m_i) - x_i = 0, \quad i = 1,2,\ldots N$

$\quad\quad\quad\quad\quad\;\; h_i(z_i,x_i,m_i) \leq 0, \quad i = 1,2,\ldots N$

Let the values of the inputs, outputs and controls which solve ML for any given value of λ be denoted by z^+, x^+ and m^+. (Again these should more correctly be denoted by $z^+(\lambda)$, $x^+(\lambda)$, $m^+(\lambda)$

because they depend upon the given value of λ but the argument is omitted for convenience.) If a value of λ, say λ^*, can be found such that the interconnection constraints are satisfied by the resulting values of z^+ and x^+, i.e. such that

$$\sum_{i=1}^{N} g_i(z_i^+, x_i^+) = 0 \qquad (3.11)$$

then $z^+ = z^*$, $x^+ = x^*$ and $m^+ = m^*$ and $L(z^+, x^+, m^+, \lambda^*) = F$. That is to say the solution of ML for $\lambda = \lambda^*$ solves GMP. For any other value of λ it may be shown that

$$L(z^+, x^+, m^+, \lambda) \geqslant F \qquad (3.12)$$

i.e. the maximum value of the Lagrangian for any given value of λ provides an upper bound on F and this upper bound will equal F if and only if there exist a (finite) $\lambda = \lambda^*$, such that (3.11) is satisfied (Whittle [21]). It follows from this that the required value of λ may be sought by minimizing $L(z^+, x^+, m^+, \lambda)$ over λ. Thus we define

$$\Phi(\lambda) = L(z^+, x^+, m^+, \lambda) \qquad (3.13)$$

and attempt to determine λ^* by solution of the problem MD.

MD: Minimize $\Phi(\lambda)$
     λ

The function $\Phi(\lambda)$ is called the dual function and MD the dual problem (i.e. the dual of GMP).

The pair of interacting problems ML and MD now replace GMP in the sense that the solution of GMP is sought by iteratively solving ML and MD in succession. The point of this problem transformation is that for a given value of λ the Lagrangian separates; thus from (3.10)

$$L(z, x, m, \lambda) = \sum_{i=1}^{N} f_i(z_i, x_i, m_i) - \lambda^T \sum_{i=1}^{N} g_i(z_i, x_i)$$

$$= \sum_{i=1}^{N} \{f_i(z_i, x_i, m_i) - \lambda^T g_i(z_i, x_i)\}$$

$$= \sum_{i=1}^{N} \ell_i(z_i, x_i, m_i, \lambda)$$

where

$$\ell_i(z_i,x_i,m_i,\lambda) = f_i(z_i,x_i,m_i) - \lambda^T g_i(z_i,x_i)$$

may be regarded as a local subsystem Lagrangian function. Thus we can replace ML by N independent problems.

ML(i): Maximize $\ell_i(z_i,x_i,m_i,\lambda)$
        z_i,x_i,m_i

    subject to $T_i(z_i,m_i) - x_i = 0$

                $h_i(z_i,x_i,m_i) \leq 0$

The information flow between MD and ML(i) at each iteration is shown in Fig 3.3.

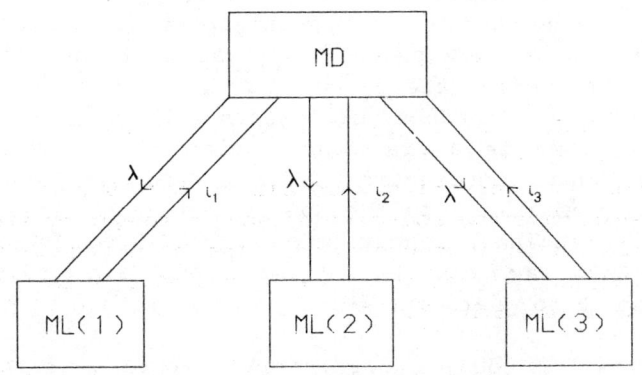

Fig 3.3 Information flow in dual coordination

In the context of interconnected production systems introduced in section 3.2.1, the dual decomposition presented above has a direct economic interpretation (Brosilow and Lasdon [4]). Note that each element of the vector

G   z
    x

on the left hand side of (3.3) is associated with just one product stream. If we assign a price to each stream and ask that each subsystem which makes or uses this stream should sell it or buy it at this price, and if the profit or loss thereby incurred is added to or subtracted from the local objective (profit) function, we find that each local objective function so modified becomes

$$f_i(z_i,x_i,m_i) - \lambda^T G_i \begin{bmatrix} z_i \\ x_i \end{bmatrix} = f_i(z_i,x_i,m_i) - \lambda^T g_i(z_i,x_i)$$

where λ is the vector of stream prices. If we allow each subsystem to choose its own values for its inputs, outputs and controls to maximize this modified profit function subject only to its local constraints, we find we have formulated the Lagrangian subproblems {ML(i)}. In this interpretation the role of the coordinator MD is to adjust the prices λ so that each subsystem finds it profitable to produce and consume just the right amount of the product streams to maintain the interconnections in balance. In the light of this interpretation dual coordination is often called 'price coordination', and because the subsystem goals $\{f_i(z_i,x_i,m_i)\}$ are modified by the terms $\{\lambda^T g_i(z_i,x_i)\}$ through which the coordinator exerts its influence, the description 'goal coordination' is also used.

We remark that this is just the interpretation of the Lagrange multipliers as shadow prices that was mentioned in chapter 1.2.1. We know from the discussion there that, under fairly weak conditions, Lagrange multipliers will exist such that the solution of GMP is a stationary point of the Lagrangian. However, in dual coordination, we are asking that the Lagrangian be actually maximized by the solution of GMP. This requires that the Lagrangian be concave in a neighbourhood of the solution of GMP. As we saw in chapter 1, simple constraint qualification is not enough to guarantee this. We will return to this point below.

Analytically, dual coordination appears rather more attractive than primal coordination; in particular, only mild assumptions of continuity of f_i, g_i, h_i are needed to ensure that the dual function $\Phi(\lambda)$ is convex, and if the solutions of ML(i) are unique the gradient of $\Phi(\lambda)$ exists and is given by

$$-\sum_{i=1}^{N} g_i(z_i^+, x_i^+).$$

Consequently, reasonably fast optimization algorithms can be used to solve MD - indeed it is possible to generate second derivative information at the local level for use in solving MD by second order optimization algorithms (Foord [9]). Because the solution of problem MD can be so tractable, dual coordination has received considerable attention in the literature which has not always reflected the limitations of the method. The obvious failing is that it is an infeasible method in the sense that the solutions z^+, x^+, m^+ derived from any value of λ other than λ^* do not satisfy the interconnection constraints (3.5), but this is a limitation only in some on-line situations. More importantly the presentation up to this point has not considered whether λ^* exists. If it does then solution of the dual problem MD will find it because $\Phi(\lambda)$ is convex, but in many problems it is observed that for the optimal solution λ^+ of MD, the corresponding solutions of the subproblems $\{ML(i)\}$ are not unique and none of the values of z^+, x^+, m^+ satisfy the constraints (3.5). In this situation the dual function is not differentiable, λ^+ cannot be identified with λ^* and the optimal value of the dual function $L(z^+, x^+, m^+, \lambda^+)$ does not equal F. Sufficient conditions on the problem GMP to ensure that this kind of failure does not occur in dual coordination, are that the functions $\{f_i\}$ are strictly concave, $\{h_i\}$ are convex and $\{T_i\}$ are linear. The effect of these very strong conditions is to ensure that, whatever the optimal values of the Lagrange multipliers, the Lagrangian for GMP will be a concave function of m, x, z (since the optimal Lagrange multipliers associated with the inequality constraints must be positive). These conditions are restrictive for practical problems; they are however sufficient conditions only, and it is quite possible that problems not satisfying these conditions can be solved using dual coordination. For example Javdan [13] has shown that certain problems with quadratic objectives and quadratic equality constraints can be solved satisfactorily using dual coordination, but it must be emphasised that in the absence of special circumstances the application of dual coordination to problems which do not satisfy the sufficient conditions given above cannot be robust and failure is highly probable (Foord [9]; Simmons and White [19]).

3.3 CONCLUSION

The broad conclusions to be drawn from the discussion given above are that in a real industrial environment, coordination by specifying production targets or product transfer prices may be difficult to implement or may not lead to the optimal policy: the aforementioned assumptions of convexity and linearity which we required to make progress with solving the mathematical programming problem GMP introduced in (a) seem restrictive and therefore the theory would appear to offer little help in guiding the solution of real problems. Whilst we believe these conclusions are largely valid in the context given, we also believe that they are too sweeping, and based on too narrow a concept of the problem, to warrant total rejection of the decentralized or hierarchical approach. More recent work has extended the validity of decentralized optimization techniques and for dynamic systems dual coordinations can provide a viable approach.

REFERENCES

[1] Athans, M. (1974) Survey of decentralized control methods. 3rd NBER/FRB Workshop on Stochastic Control, Washington D. C.

[2] Bailey, F. N. (1976) Decision processes in organisations. in Large-scale dlynamical systems (ed. R. Saeks). Point Lobos Press: California.

[3] Benders, J. F. (1962) Partitioning procedures for solving mixed- variable programming problems. Numerisch Mathematik, 4, 238-252.

[4] Brosilow, C. B. and Lasdon, L. S. (1965) A two level optimization technique for recycle processes. Proceedings of the A.I.Ch.E.-Ind. Chem. Eng. Joint Meeting, London, 4:75-4:83.

[5] Dantzig, G. B. and Wolfe, P. (1960) Decomposition principle for linear programs. Operations Research 8, 101-111.

[6] Drew, S. A. W. (1975) A study in the application of large scale control methods to a practical industrial problem. Proceedings of the 6th IFAC World Congress, Boston, 3.1:1-3.1:9

[7] Findeisen W. (1974) Wielopoziomowe układy sterowania. Panstwowe Wydawnictwo Naukowe: Warsaw

[8] Findeisen, W. (1974) Control and coordination in multilevel systems. 2nd Polish-Italian Conference on Applications of Systems Theory to Economy, Management and Technology, Pugnoschiuso

[9] Foord, A. G. (1974) On-line optimization of a petrochemical complex. Ph.D. Thesis: Cambridge University

[10] Geoffrion, A. M. (1970) Elements of large-scale programming. Management Sci. Theory, 16,652-691

[11] Himmelblau, D. M. (ed) (1973) Decomposition of large-scale problems. North Holland Publishing Co.: Amsterdam

[12] Ho, Y. C. and Chu, K. C. (1974) Information structure in dynamic multi-person control problems. Automatica, 10, 341-351

[13] Javdan, M. R. (1976) Extension of dual coordination to a

class of non-linear systems. Int. J. Control, 24, 551-572

[14] Kulikowski, R., Krus, L., Manczak, K. and Straszak, A. (1975) Optimization and control problems in large-scale systems. Proc. of the 6th IFAC World Congress, Boston, 19.1:1 - 19.1:13

[15] Lasdon, L. S. (1970) Optimization theory for large systems. MacMillan: London

[16] Mesarovic, M. D., Macko, D. and Takahara, Y. (1970) Theory of hierarchical multilevel systems. Academic Press: New York

[17] Rosen, J. E. (1964) Primal partition programming for block diagonal matrices. Numerische Mathematik, 6, 250-260

[18] Sandell, N. R. Varaiya, P. P. aand Athans, M. (1975) A surgey of decentralized control methods for large-scale systems. Proceedings Engineering Foundation Conference on Systems Engineering for Power Systems, Henniker, New Hampshire

[19] Simmons, M. D. and White, G. W. T. (1977) Analysis of complex systems. Trans. R. Soc. Lind. A, 287, 405-423

[20] Singh, M. and Titli, P. (1978) Systems decomposition, optimization and control. Pergamon: London

[21] Whittle, P. (1971) Optimization under constraints. Wiley: London

[22] Wilson, I. D. (1977) Decentralized control based on problem decomposition. Ph. D. Thesis: Cambridge

Chapter 4
Introduction to linear programming
L.H. Campbell

4.1 INTRODUCTION

A linear program is an optimization problem with a linear objective function and linear constraints. Ever since the discovery in 1947 by G.B. Dantzig [2] of the simplex method for the solution of such problems, linear programs have attracted much interest in operational research. Many applications have involved inherently linear functions; in other cases, it has been attractive to make linear approximations to non-linear programs. Well developed computer codes are available for the solution of very large linear programs.

Although the theory of linear programming has tended to develop without reference to the general theory of static optimization, there is nevertheless a close relationship between the two. This chapter indicates how the special features of linear programs may be exploited in the simplex algorithm for their solution, and how this method is related to the duality theory for such optimization problems. Some operational considerations in the building and use of linear programming models are outlined.

4.2 FORM OF A LINEAR PROGRAM

A linear program may always be written in the following form:

$$\text{maximize} \quad p = p_0 + \sum_{j=1}^{n} c_j x_j$$

$$\text{subject to} \quad \sum_{j=1}^{n} a_{ij} x_j = b_i, \quad i = 1, 2, \ldots, m$$

$$x_j \geqslant 0, \quad j = 1, 2, \ldots, n.$$

78 Linear programming

Here, $x_1,\ldots,x_n$ are the variables whose values are to be chosen to maximize the function p. The coefficients $\{c_j\}$ and $\{a_{ij}\}$ and the terms p_0 and $\{b_i\}$ are all constants of known value. In vector notation we may write:

maximize $\quad p = p_0 + c^T x$

subject to $Ax = b$

$x \geq 0$,

where x is a (column) n-vector of variables,
c is a (column) n-vector of coefficients,
b is a (column) m-vector of constants,
A is an m×n matrix of coefficients,
p_0 is a scalar.

If the equations $Ax = b$ have either exactly one solution, or no solutions at all, then no optimization is necessary. If some of these constraints are redundant, they may be eliminated by writing them as linear combinations of the other constraints. Hence, without loss of generality, we can assume that $n \geq m$ and rank(A) = m. These conditions are needed only for ease of exposition. In practice, one may use a linear programming code to solve problems violating one or both conditions. The second, that A be of full rank, is sometimes difficult to check in a real-life problem anyway, but the simplex algorithm (section 4.4) copes adequately with redundant constraints. In the case $n < m$, one would normally solve the dual problem (section 4.8).

The problem of linear programming is to determine whether solutions exist to a given linear program and, if so, to find a solution which yields an optimal value of the objective. To write a problem in the form given in this section, one may need to introduce slack variables into inequality constraints to produce equalities, and one may need to re-write some variables so that all are constrained to be non-negative.

4.3 BASIC SOLUTIONS

Suppose we write $a^{(j)}$ for the jth column of A. Then the equality constraints may be written

$$\sum_{j=1}^{n} a^{(j)} x_j = b.$$

Choose m linearly independent columns $\{a^{(i_1)}, a^{(i_2)},\ldots,a^{(i_m)}\}$, and let B be the matrix formed from them: that is

$$B = [a^{(i_1)}, a^{(i_2)}, \ldots, a^{(i_m)}].$$

Then B is $m \times m$ and non-singular. Premultiplying $Ax = b$ by B^{-1} yields

$$B^{-1}Ax = B^{-1}b,$$

in which the columns $i_1, i_2, \ldots, i_m$ of the product $B^{-1}A$ form the $m \times m$ unit matrix. Hence a solution of $Ax = b$ is

$$(x_{i_1}, x_{i_2}, \ldots, x_{i_m})^T = B^{-1}b,$$

$$x_j = 0 \text{ for all other } j.$$

This is called a basic solution of $Ax = b$. The set of indices $S = \{i_1, i_2, \ldots, i_m\}$ is called a basis. Note that the order of the entries in S is significant. The corresponding variables $x_{i_1}, x_{i_2}, \ldots, x_{i_m}$ are called basic variables; the remaining variables are called non-basic. If, in addition, the constraints $x \geq 0$ are satisfied, the solution is called a basic feasible solution.

Example: Consider the equations

$$\begin{bmatrix} -1 & 1 & 1 & 0 \\ 1 & 1 & 0 & 1 \end{bmatrix} \begin{bmatrix} x_1 \\ x_2 \\ x_3 \\ x_4 \end{bmatrix} = \begin{bmatrix} 1 \\ 2 \end{bmatrix}.$$

Choose $S = \{3, 4\}$. Then $B = I$, and the corresponding solution is $x_1 = 0$, $x_2 = 0$ and

$$\begin{bmatrix} x_3 \\ x_4 \end{bmatrix} = \begin{bmatrix} 1 & 0 \\ 0 & 1 \end{bmatrix} \begin{bmatrix} 1 \\ 2 \end{bmatrix} = \begin{bmatrix} 1 \\ 2 \end{bmatrix}.$$

This is a basic feasible solution. Similarly, choose $S = \{1, 4\}$. Then

$$B = \begin{bmatrix} -1 & 0 \\ 1 & 1 \end{bmatrix}.$$

and $B^{-1} = B$. The corresponding solution has $x_2 = 0$, $x_3 = 0$, $x_1 = -1$, $x_4 = 3$. This is not feasible. In this case,

$$B^{-1}A = \begin{bmatrix} -1 & 0 \\ 1 & 1 \end{bmatrix} \begin{bmatrix} -1 & 1 & 1 & 0 \\ 1 & 1 & 0 & 1 \end{bmatrix},$$

$$= \begin{bmatrix} 1 & -1 & -1 & 0 \\ 0 & 2 & 1 & 1 \end{bmatrix}.$$

Suppose we wish to replace x_4 by x_2 amongst the basic variables. From the column of current coefficients of x_2, namely,

$$\begin{bmatrix} -1 \\ 2 \end{bmatrix},$$

we generate (as described in detail in the next section) the elementary matrix

$$\begin{bmatrix} 1 & 1/2 \\ 0 & 1/2 \end{bmatrix},$$

and premultiply B^{-1} by this matrix to get

$$\begin{bmatrix} 1 & 1/2 \\ 0 & 1/2 \end{bmatrix} \begin{bmatrix} -1 & 0 \\ 1 & 1 \end{bmatrix} = \begin{bmatrix} -1/2 & 1/2 \\ 1/2 & 1/2 \end{bmatrix}.$$

The reader may verify that this is the inverse of

$$\begin{bmatrix} -1 & 1 \\ 1 & 1 \end{bmatrix},$$

and hence is the inverse basis matrix corresponding to $S = \{1,2\}$. Thus we may move from one basis to another without re-inverting the matrix B. The corresponding solution is $x_3 = 0$, $x_4 = 0$, $x_1 = 1/2$, $x_2 = 3/2$.

4.4 THE REVISED SIMPLEX METHOD

In this section, we describe an algorithm, called the revised simplex method, which may be used to solve any linear program presented in the form of section 4.2: that is, with equality constraints and all variables constrained to be non-negative. The details of why the algorithm works are indicated in subsequent sections.

Step 0. (Crashing or Phase 1)

Choose an initial basis $S = \{i_1, i_2, \ldots, i_m\}$ and corresponding basis matrix B such that $B^{-1}b \geq 0$ (i.e. the corresponding basic solution is feasible). Set

$$\bar{b} = B^{-1}b \quad \text{(m rows)}$$

$$\pi^T = (c_{i_1}, c_{i_2}, \ldots, c_{i_m}) B^{-1} \quad \text{(m columns)}$$

$$\bar{p} = \pi^T b + p_0$$

$$\bar{S} = \{j : 1 \leq j \leq n, j \notin S\}$$

($\bar{S}$ is the set of indices of the non-basic variables).

Step 1. (Pricing out the columns)

Recall that $a^{(j)}$ is the jth column of A.

For each $j \in \bar{S}$, compute $d_j = \pi^T a^{(j)} - c_j$.

Let s be such that $d_s = \min_{j \in \bar{S}} \{d_j\}$.

If $d_s \geq 0$, stop: the current basic solution is <u>optimal</u>.

Otherwise $d_s < 0$: go to step 2.

Step 2. (Choosing the pivot row)

Compute $\bar{a}^{(s)} = B^{-1} a^{(s)} = (\bar{a}_{1s}, \bar{a}_{2s}, \ldots, \bar{a}_{ms})^T$, say.

If $\bar{a}^{(s)} \leq 0$, stop: the objective p <u>increases without bound</u> in the feasible region.

Otherwise, choose k such that

$$\frac{\bar{b}_k}{\bar{a}_{ks}} = \min_{\substack{1 \leq i \leq m \\ \bar{a}_{is} > 0}} \{\frac{\bar{b}_i}{\bar{a}_{is}}\}.$$

If there is not a unique choice of k, an appropriate tie-breaking rule is required to guarantee convergence of the algorithm. The details of such a rule are not important for most practical purposes.

Let t be the kth index in S.

Step 3. (Pivot step)

Replace t by s at the kth position in S .

Replace s by t in $\bar{S}$.

Premultiply the $(m+1) \times (m+1)$ matrix

$$\begin{bmatrix} B^{-1} & \bar{b} \\ \pi^T & \bar{p} \end{bmatrix}$$

by the matrix formed from the $(m+1) \times (m+1)$ identity matrix by replacing the kth column by η, where

$$\eta_i = \begin{cases} \dfrac{-\bar{a}_{is}}{\bar{a}_{ks}}, & i=1,2,\ldots,m,\ i \neq k \\[6pt] \dfrac{1}{\bar{a}_{ks}}, & i=k \\[6pt] \dfrac{-d_s}{\bar{a}_{ks}}, & i=m+1 \end{cases}$$

(Except for $i=k$ and $i=m+1$, these values are just the entries of $\bar{a}^{(s)}$ divided by $-\bar{a}_{ks}$.)

This matrix product provides the new values for: B^{-1}, the inverse basis matrix; $\bar{b}$, the values of the (now) basic variables; $\bar{p}$, the objective value; and π^T, the so-called simplex multipliers.

Go to step 1.

Example

$$\text{maximize} \quad p = x_1 + 3x_2$$

$$\text{subject to} \quad -x_1 + x_2 + x_3 \quad\ = 1$$

$$\phantom{\text{subject to}} \quad\ \ x_1 + x_2 \quad\ + x_4 = 2$$

$$x_j \geq 0, \quad j = 1,2,3,4.$$

The constraints and objective function may be written

$$\begin{bmatrix} -1 & 1 & 1 & 0 \\ 1 & 1 & 0 & 1 \\ 1 & 3 & 0 & 0 \end{bmatrix} \begin{bmatrix} x_1 \\ x_2 \\ x_3 \\ x_4 \end{bmatrix} = \begin{bmatrix} 1 \\ 2 \\ p \end{bmatrix}.$$

Step 0. Let $S = \{3,4\}$; $\bar{S} = \{1,2\}$. Then $B^{-1} = I$ and

$$\bar{b} = \begin{matrix} 1 \\ 2 \end{matrix} \qquad \pi^T = (0,0) \qquad \bar{p} = 0.$$

Step 1. $d_1 = -1$; $d_2 = -3$, hence $s = 2$ with $d_2 = -3 < 0$.

Step 2. $\bar{a}^{(2)} = (1,1)^T$, $\bar{b} = (1,2)^T$. Hence $k = 1$ and $t = 3$.

Step 3. $S = \{2,4\}$, $\bar{S} = \{1,3\}$.

$$\begin{bmatrix} 1 & 0 & 0 \\ -1 & 1 & 0 \\ 3 & 0 & 1 \end{bmatrix} \begin{bmatrix} 1 & 0 & 1 \\ 0 & 1 & 2 \\ 0 & 0 & 0 \end{bmatrix} = \begin{bmatrix} 1 & 0 & 1 \\ -1 & 1 & 1 \\ 3 & 0 & 3 \end{bmatrix}$$

so that the new solution is $x_2 = 1$, $x_4 = 1$, $x_1 = x_3 = 0$, which yields an objective value of 3.

$$\pi^T = (3,0), \qquad B^{-1} = \begin{bmatrix} 1 & 0 \\ -1 & 1 \end{bmatrix}.$$

Step 1. $d_1 = -4$; $d_3 = 3$, hence $s = 1$ with $d_s = -4 < 0$.

Step 2. $a^{(1)} = (-1,2)^T$, $\bar{b} = (1,1)^T$, hence $k = 2$ (the only candidate) and $t = 4$.

Step 3. $S = \{2,1\}$; $\bar{S} = \{4,3\}$

$$\begin{bmatrix} 1 & 1/2 & 0 \\ 0 & 1/2 & 0 \\ 0 & 2 & 1 \end{bmatrix} \begin{bmatrix} 1 & 0 & 1 \\ -1 & 1 & 1 \\ 3 & 0 & 3 \end{bmatrix} = \begin{bmatrix} 1/2 & 1/2 & 3/2 \\ -1/2 & 1/2 & 1/2 \\ 1 & 2 & 5 \end{bmatrix}$$

Solution is now $x_2 = 3/2$, $x_1 = 1/2$, $x_4 = x_3 = 0$, $p = 4$.

Step 1. $d_4 = 2$; $d_3 = 1$.

Minimum is $d_3 = 1 > 0$.

Therefore, stop: the optimal solution is $x_1^* = 1/2$, $x_2^* = 3/2$, $x_3^* = x_4^* = 0$, with maximal objective value $p^* = 4$.

4.3 CONVERGENCE OF THE SIMPLEX ALGORITHM

In this section, some of the details of why the simplex algorithm works are sketched. The detailed proofs, where they do not lead to new understanding, are omitted.

The simplex algorithm proceeds by stepping from one basic feasible solution to another, changing only one basis entry at a time, until an optimal solution is reached. The reader should verify that step 3, the pivot step, does indeed do what is claimed for it: that is, that it does produce the values of B^{-1}, $\bar{b}$ and $\bar{p}$ corresponding to the new basis. Consider a particular basic feasible solution and re-order the variables, if necessary, so that those that are basic appear first in the order in which their indices occur in the basis. Let $x^{(B)}$ be the m-vector of basic variables; let $x^{(D)}$ be the vector of the (n-m) remaining non-basic variables. Partition the objective and constraints as follows:

$$\text{maximize} \quad p = p_0 + c^{(B)T} x^{(B)} + c^{(D)T} x^{(D)}$$

$$\text{subject to} \quad [B \ D] \begin{bmatrix} x^{(B)} \\ x^{(D)} \end{bmatrix} = b$$

$$x^{(B)} \geq 0, \ x^{(D)} \geq 0,$$

where $c^{(B)}$ and $c^{(D)}$ are the first m and the last (n-m) entries respectively in the n-vector c, and B and D consist of the first m and the last (n-m) columns, respectively, of A.

Premultiplying the constraints by B^{-1} and rearranging gives:

$$x^{(B)} = B^{-1}b - B^{-1}Dx^{(D)},$$

whence

$$p = p_0 + c^{(B)T}B^{-1}b - (c^{(B)T}B^{-1}D - c^{(D)T})x^{(D)}. \qquad (4.1)$$

The reader should verify that, throughout the simplex algorithm, the identity

$$\pi^T = c^{(B)T}B^{-1}$$

always holds and hence that the vector

$$c^{(B)T}B^{-1}D - c^{(D)T}$$

is just the row of coefficients d_j, $j \epsilon \bar{S}$, of the simplex algorithm.

If the simplex algorithm is to succeed in finding an optimal solution, it is necessary that there should be at least one such solution which is basic. That this is always so, at least in the case where the set of feasible solutions is bounded, may be seen as follows. (The unbounded case is only slightly more difficult.) Suppose one has a solution in which $x_1, \ldots, x_r$ are all positive, with $r > m$. Write the constraints and objective with $x_1, \ldots, x_m$ basic (re-ordered if necessary) and consider varying the value of x_r to $x_r + \delta$. If δ is of opposite sign to d_r (either sign if $d_r = 0$) then the objective value will not decrease. By making the magnitude of δ as large as possible, while retaining feasibility, either x_r will become zero or one of the basic variables will become zero; in the latter case, the basis should be written with r in place of the index of the variable which first becomes zero. In either case, the result is that the number of non-zero values in the solution has been reduced by (at least) one, while the objective value has not decreased. This process may be continued until there are no more than m non-zero values, which corresponds to a basic feasible solution.

The algorithm proceeds by selecting in step 1 a variable x_s whose value, presently 0, when increased will lead to an increase in objective value. Step 2 determines the largest increase possible in the value of x_s while retaining feasibility. Step 3 then enters s in the basis and removes the index of the variable which has become 0. With an appropriate tie-breaking rule in step 2, it may be guaranteed that no basis ever recurs. (For problems in which the objective value always strictly increases in step 3, it is clear that no basis can recur: difficulties can only arise when a pivot step is performed in which the objective value is unchanged.) As there is only a finite number of possible bases, the algorithm will terminate in a finite number of iterations.

The stopping condition in step 1 is just that

$$c^{(B)T}B^{-1}D - c^{(D)T} \geqslant 0.$$

With $x^{(D)} = 0$, then if the solution is to satisfy the constraints,

$$x^{(B)} = B^{-1}b$$

and hence

$$p = p_0 + c^{(B)T} B^{-1} b.$$

Any other solution, therefore, must have at least one of the entries of $x^{(D)}$ strictly positive; but, because of (4.1), the stopping condition says that any such increase in the value of an entry in $x^{(D)}$ away from 0 will not increase the value of the objective. Hence an optimal solution has been reached. A similar argument may be applied to the stopping condition in step 2: in this case, x_s may be increased in value without bound while making the objective arbitrarily large.

4.6 CHOOSING AN INITIAL BASIC FEASIBLE SOLUTION

Step 0 of the simplex algorithm requires the choice of a basic feasible solution. If there is no other information available, one can proceed as follows. First, write the constraints with $b \geq 0$, by changing the signs of the constraint equations as necessary. Then solve the extended problem

maximize $\quad p' = -1^T x'$

subject to $Ax + Ix' = b$

$x, x' \geq 0,$

where x' is an m-vector of so-called artificial variables, 1 is a vector of units and I is the m×m identity matrix.

This new linear program may be started with x' as the initial vector of basic variables. If this problem has optimal value zero, then the final solution provides an initial basic feasible solution for the original problem. If the new problem has negative optimal objective value, however, then no feasible solutions exist for the original problem.

Solving the above linear program is called phase 1 of the simplex algorithm. In practical problems, however, there is usually some information on which variables will be positive in an optimal solution. Computer codes will often allow the user to specify a list of such variables, which will then be used in an initial basis: artificial variables are then needed only to make up a feasible solution of full rank. Such advanced starts will often significantly shorten the computing time required to solve a big problem.

4.7 POST-OPTIMALITY ANALYSIS

In practical problems, it is often the case that one is uncertain as to the exact values of the various constants in a linear program, or one may know that these values in fact will change with time. It is of interest therefore to know something of how the optimal solution will vary in response to these changes. Sensitivity information of this kind is readily available in linear programming. In particular, the ranges within which the right-hand-side values and the objective-function coefficients may lie, while still retaining the same optimal basis, are often provided automatically by computer packages.

4.7.1 Right-hand-side ranges

Suppose the right hand side value, b_i, of the ith constraint were to change to $b_i + \theta$. Then the vector b becomes $b + \delta^{(i)}\theta$, where $\delta^{(i)}$ is the ith column of the m×m unit matrix. The coefficients d_j, $j \in \bar{S}$, are unaffected, so the present optimal basis will remain optimal provided only that the corresponding solution remains feasible, that is while

$$B^{-1}(b + \delta^{(i)}\theta) \geq 0.$$

This set of m inequalities provides simple bounds for θ.

Example: In the example of Section 4.4, the right-hand-side range for the first constraint may be calculated as follows. For the present optimal basis,

$$B^{-1}b + B^{-1}\delta^{(1)}\theta = \begin{bmatrix} 3/2 \\ 1/2 \end{bmatrix} + \begin{bmatrix} 1/2 \\ -1/2 \end{bmatrix} \theta$$

and this vector is non-negative provided $-3 \leq \theta \leq 1$. Hence the right-hand-side value, presently 1, may vary between -2 and 2 and the basis $S = \{2, 1\}$ will remain optimal.

4.7.2 Objective-coefficient ranges

This is equivalent to right-hand-side ranging for the dual problem (section 4.8). The analysis, however, leads to an interpretation of the values d_j computed in step 1 of the simplex method.

Suppose the objective coefficient of a non-basic variable x_j

becomes $c_j + \theta$. The optimality condition of step 1 of the simplex method is satisfied provided

$$\pi^T a^{(j)} - (c_j + \theta) \geq 0,$$

or, equivalently, $\theta \leq d_j$. It is clear why there should be no lower bound on θ: it is currently unprofitable to make x_j positive and no decrease in the unit value of this variable will ever make it more attractive. The value d_j is just the minimum amount by which the unit profit on this variable must be increased before it becomes attractive to undertake the activity j.

If the objective coefficient of the kth basic variable changes to $c_{i_k} + \theta$, then the values of the simplex multipliers π are changed and the optimality condition of step 1 becomes

$$(c^{(B)T} + \delta^{(k)T}\theta)B^{-1}D - c^{(D)T} \geq 0,$$

or

$$d_j + \delta^{(k)T}B^{-1}a^{(j)}\theta \geq 0, \text{ for all } j \in \bar{S}.$$

These n-m inequalities provide simple bounds on θ.

Example: In the example of Section 4.4, suppose the coefficient of x_1 becomes $1 + \theta$. The product

$$\delta^{(2)T}B^{-1} = (-1/2, 1/2),$$

(that is, just the second row of B^{-1}) and hence the inequalities on θ are just that $2 + (1/2)\theta$ and $1 - (1/2)\theta$ should both be non-negative. Thus $-4 \leq \theta \leq 2$ and the objective-coefficient range is -3 to 3, inclusive.

4.7.3 Other changes

The above analyses provide for the alteration of only one coefficient at a time. Exactly equivalent calculations may be done when several coefficients change in unison. This is generally called parametric programming: for an example, see Beale [1], ch.6. Changes to the entries in the matrix A may also be handled, although computer packages often do not provide facilities automatically for doing so: for a full discussion, see Rao [5], pp.177-184. It should be clear that it is never necessary to re-solve a linear program from the very beginning. The optimal basis provides a starting point for the solution to a modified version of the problem and artificial variables need only be added to resolve infeasibilities.

4.8 DUALITY IN LINEAR PROGRAMMING

A symmetric duality theory for linear programming may be obtained by considering an inequality-constrained problem and noting the special role of the slack variables. Hence for the purposes of this discussion, we consider the special linear program

maximize $\quad p = c^T x$

subject to $Ax \leqslant b$

$\quad\quad\quad\quad x \geqslant 0$

Adding an m-vector of slack variables z allows the constraints to be written

$\quad Ax + z = b$

$\quad x, z \geqslant 0.$

Introduce Lagrange multipliers y, and define for each y

$\quad w(y) = \max_{x,z \geqslant 0} \{c^T x + y^T(b-Ax-z)\}.$

Then

$\quad w(y) = \max_{x,z \geqslant 0} \{c^T x + y^T(b-Ax) - y^T z\},\quad\quad (4.2)$

$\quad\quad\quad = \max_{x,z \geqslant 0} \{y^T b + (c^T - y^T A)x - y^T z\}.\quad\quad (4.3)$

The Lagrangian dual problem is to minimize $w(y)$ over y. Considering the maximization over z in (4.2),

$\quad \max\{-y^T z\} = +\infty$

if any component of y is negative. Hence $w(y)$ has its minimum at some $y \geqslant 0$. In (4.3), maximizing over x,

$\quad \max\{(c^T - y^T A)x\} = +\infty$

unless $c^T - y^T A \leq 0$. If $(c^T - y^T A) \leq 0$, $y \geq 0$, the maximization over $x, z \geq 0$ is performed by arranging that x, z satisfy

$$(c^T - y^T A) x = 0$$

$$y^T z = 0.$$

Then the form in y to be minimized is just $y^T b$. Thus the dual problem may be written

minimize $\quad w = y^T b$

subject to $\quad y^T A \geq c^T$

$$y \geq 0.$$

Similar forms of dual problem, varying in the exact form of their constraints, can be obtained for other forms of the primal linear program. The reader should investigate these, and also verify that the dual of the dual is the primal, in each case. This leads one to expect that solving one of a primal-dual pair should lead directly to a solution of the other. We now investigate this point.

Consider the primal-dual pair

P: maximize $\quad p = c^T x$

subject to $Ax \leq b$

$$x \geq 0$$

D: minimize $\quad w = y^T b$

subject to $\quad y^T A \geq c^T$

$$y \geq 0.$$

The following lemma is sometimes called the Weak Duality theorem of linear programming.

Linear programming

Lemma

If x,y are feasible for P,D respectively, then $c^T x \leq y^T b$.

Proof

Feasibility of x,y for P,D implies

$$c^T x \leq (y^T A) x = y^T (Ax) \leq y^T b.$$

This result tells us that any feasible solution y for D provides an upper bound $b^T y$ for the optimal primal objective value, and conversely.

Duality theorem

(i) If P and D both have feasible solutions, they both have optimal solutions, with equal objective function values.

(ii) If P is unbounded, D is infeasible.

(iii) if D is unbounded, P is infeasible.

The proof of (i) follows from Lagrangian duality theory (chapter $.$), since we have seen already that D is equivalent to the Lagrangian dual of P. An alternative proof is provided by the simplex algorithm itself, and this is discussed in the next section. The proof of (ii) and (iii) follows directly from the lemma. Note that it is possible for both P an D to be infeasible.

Complementary slackness conditions

Let x,y be *optimal* solutions to P, D respectively. Then

(i) if $x_j > 0$, then $(y^T A)_j = c_j$, $j = 1, 2, \ldots, n$;

(ii) if $y_i > 0$, then $(Ax)_i = b_i$, $i = 1, 2, \ldots, m$;

(iii) if $(y^T A)_j > c_j$, then $x_j = 0$, $j = 1, 2, \ldots, n$;

(iv) if $(Ax)_i < b_i$, then $y_i = 0$, $i = 1, 2, \ldots, m$.

Proof

The objective values in P,D respectively of x,y must be equal, by (i) of the duality theorem. This will be so only if both inequalities in the proof of the lemma are equalities. Since $x,y \geqslant 0$, this can happen only if (i) and (ii) hold; (iii) and (iv) are just re-statements of (i) and (ii) in terms of the slack variables for P and D.

Partial converse

If x,y are feasible for P,D respectively and satisfy the complementary slackness conditions, then x,y are optimal for P,D.

Proof

Since x,y are feasible for P,D the weak duality theorem tells us that $b^T y$, $c^T x$ are an upper and lower bound respectively for the optimal primal and dual objective values. Since x and y satisfy the complementary slackness conditions, they achieve these bounds.

One of the most elegant features of the simplex algorithm, and the theory of linear programming in general, is the way that the duality properties, derived here by essentially Lagrangian methods, are manifested in the algebraic manipulations of the algorithm.

4.9 DUALITY AND THE SIMPLEX METHOD

Suppose we partition the equality constrained primal program as in section 4.5:

$$\text{maximize} \quad p = p_0 + c^{(B)T} x^{(B)} + c^{(D)T} x^{(D)}$$

$$\text{subject to} \quad [B \ D] \begin{bmatrix} x^{(B)} \\ x^{(D)} \end{bmatrix} = b$$

$$x^{(B)} \geqslant 0, \quad x^{(D)} \geqslant 0.$$

The dual problem is then

minimize $\quad w = p_0 + y^T b$

subject to $y^T B \geqslant c^{(B)T}$,

$\quad\quad\quad\quad y^T D \geqslant c^{(D)T}$,

with y unrestricted in sign.

Consider the vector π of simplex multipliers in the simplex algorithm. It satisfies

$$\pi^T = c^{(B)T} B^{-1},$$

or

$$\pi^T B = c^{(B)T}.$$

Hence

$$(\pi^T B - c^{(B)T}) x^{(B)} = 0.$$

Moreover, since $x^{(D)} = 0$, we have

$$(\pi^T D - c^{(D)T}) x^{(D)} = 0.$$

Thus at each stage of the simplex algorithm the vectors x, π satisfy the complementary slackness conditions. The optimality condition of the algorithm is just that

$$\pi^T D - c^{(D)T} \geqslant 0,$$

that is that π is feasible for the dual. If this condition holds, then x, π are feasible for primal and dual respectively, and satisfy the complementary slackness conditions, and so are optimal for primal and dual respectively.

The simplex algorithm automatically provides the solutions to both primal and dual programs. This means in particular that we can choose which to work with before starting. If the primal has more constraints than variables, one should solve the dual, for which the reverse obtains, and for which the basis is of smaller order.

4.10 OPERATIONAL CONSIDERATIONS

Much experience in the use of linear programming over a wide range of modelling applications has been built up since the 1950's. A great deal of success has been achieved in commercial applications: typically, the optimal solution may be 2 to 10% superior to manually generated solutions. This has led to much investment in the development of fast and reliable computer codes. The benefit to be gained from an LP model, however, depends also on the quality of the modelling effort that goes into its construction. It is often attractive to make linear approximations to non-linearities and to treat as deterministic some probabilistic data, but these approximations must be recognised in the interpretation and communication of the results: at the very least an extensive post-optimality analysis is a necessary part of a commercial application. Multiple objectives are not uncommon in practice. Various modelling techniques are available to handle these: for a modern, engineering and industrial view of one such, goal programming, see Ignizio [3].

The computing technology for handling linear programs is now well developed. It has been found that, except for very small problems, it is efficient to present the problem data to a computer program called a matrix generator which will write a file in the input format of an appropriate LP code. Matrix generators are often purpose-built, especially for problems which are to be solved many times, but more general packages are also available. There also exist matrix generator generators which construct a matrix generator, usually in Fortran, from a suitable problem statement.

Commercial LP codes are designed to solve real problems as quickly as possible and to allow a wide range of facilities and control. As solution times depend critically on the order of the basis matrix (the solution time increases roughly as the cube of the number of rows), the packages normally exploit special structures within the problem. For example, simple upper bounds may be handled implicitly by a straightforward modification to the simplex algorithm. (Simple lower bounds are handled by a change of origin.) LP codes are written for specific computers. Some commercial examples are MPSX for IBM 360/370, FMPS/SPRINT and SCICONIC for UNIVAC 1100, XDLA for ICL 1900, APEX II/III for CDC 6600/7600, and LAMPS and MINILP for a range of mini-computers.

Although the model builder will necessarily be involved in the interpretation of solutions, much of the process of translating output from LP codes into more easily useable form may be automated. Report writing programs are often purpose-written for a specific application. They are particularly useful in cases where a matrix generator has undertaken some problem reduction, through substitution of variables or removal of redundant constraints, before presenting the data to an LP code. In operational applications, in which a model will be used many times, much programming effort may be put into producing suitable matrix generator/report writer pairs.

REFERENCES

[1] Beale, E.M.L. (1968), Mathematical Programming in Practice, Pitman, London.

[2] Dantzig, G.B. (1963), Linear programming and extensions. Princeton University Press, Princeton.

[3] Ignizio, James P. (1976), Goal Programming and Extensions, D.C. Heath, Lexington, U.S.A.

[4] Lasdon, L.S. (1970), Optimization theory for large systems, Macmillan, New York.
Includes a clear exposition of the revised simplex method and the various modifications to it.

[5] Rao, S.S. (1978), Optimization: theory and applications. Wiley Eastern, New Delhi.
Written as a textbook for engineeering undergraduates, this book will become a standard reference. Not only does it contain a clear description of optimization techniques and their applications in engineering, it is also extremely cheap. (It is subsidised by the Indian Government.)

[6] Whittle, P. (1971), Optimization under constraints Wiley-Interscience, London.
Chapter 4, in particular, shows the connections between linear programmes and more general optimization problems.

[7] Williams, H.P. (1978), Model Building in Mathematical Programming, Wiley-Interscience, Chichester (U.K.)
A valuable book containing examples of how one may build linear programming models and present them to a computer code for solution.

Chapter 5
Decomposition in linear programming
L.H. Campbell

5.1 INTRODUCTION

Very large linear programs sometimes arise in the consideration of management problems. A company-wide model for a large organization, for example, may involve many thousands of variables and thousands of operating constraints. (The largest problems reported as having been solved include of the order of 10^4 constraints and 10^6 variables.) To find solutions to such large linear programs, the special structure of the problem needs to be exploited. Considerable savings in computer time and store may be made by suitable packing of constraint coefficients, most of which will be zero in large problems, and by modifying the simplex algorithm implicitly to account for simple constraints such as upper bounds on the variables. Commercially available computer codes generally take these features into account. In very large problems, however, it may be desirable to exploit the more detailed structure of the problem.

Most large linear programs in fact arise through the aggregation of smaller programs. For example, a company with several operating units may use sub-models of each unit to assist in decision-making within the unit but may wish to combine these sub-models into a model of the company as a whole. Optimization within each unit without reference to the others can only lead, in general, to sub-optimization of operations as a whole, so there are advantages to be gained, theoretically at least, in combining and co-ordinating sub-models. The optimal solution of the large problem will bear some relationship to the solutions of the sub-problems and it is the process of exploiting this relationship which is known as decomposition. It is not possible, of course, to decouple the sub-problems from one another entirely, so that each sub-program, slightly modified each time, may need to be solved several times before an optimal solution to the complete linear program is found. This increased computing effort, however, must be set against the benefits to be gained from the necessity of solving much smaller linear programs at each stage.

One method of decomposition, called the Dantzig-Wolfe decomposition algorithm, is described below. This method achieves coordination between the sub-problems by a "pricing" procedure in which the values of the dual variables of the complete linear program are used as internal prices of the constraints common to all sub-problems. A full description of the algorithm is given in Dantzig [1].

5.2 BLOCK ANGULAR STRUCTURE

Consider a linear program of the form:

$$\text{maximize} \quad p = c_0^T x_0 + c_1^T x_1 + \ldots + c_\ell^T x_\ell$$

subject to

$$\begin{bmatrix} A_0 & A_1 & A_2 & \cdots & A_\ell \\ & B_1 & & & \\ & & B_2 & & \\ & & & & \\ & & & & B_\ell \end{bmatrix} \begin{bmatrix} x_0 \\ x_1 \\ x_2 \\ \\ x_\ell \end{bmatrix} = \begin{bmatrix} b_0 \\ b_1 \\ b_2 \\ \\ b_\ell \end{bmatrix}$$

and all variables non-negative, where

$x_0, x_1, \ldots, x_\ell$ are <u>vectors</u> (of variables) of lengths $n_0, n_1, \ldots, n_\ell$ respectively,

$A_0, A_1, \ldots, A_\ell$ are matrices (of constants) of order $(m \times n_0), (m \times n_1), \ldots, (m \times n_\ell)$ respectively,

each $B_k, k = 1, \ldots, \ell$, is a matrix (of constants) of order $(m_k \times n_k)$,

$b_0, b_1, \ldots, b_\ell$ are vectors (of constants) of lengths $m, m_1, \ldots, m_\ell$, respectively,

$c_0, c_1, \ldots, c_\ell$ are vectors (of constants) of lengths $n_0, n_1, \ldots n_\ell$ respectively,

and spaces in the large matrix of coefficients represent zero-matrices. Such a linear program is said to have a block angular structure.

Such a program may arise in an organisation with ℓ operating units. The vector x_k, $k=1,\ldots,\ell$, represents the decision variables of the kth operating unit and the relations

$$B_k \, x_k = b_k$$

represent the local constraints on the kth unit. The constraints

$$\sum_{k=0}^{\ell} A_k \, x_k = b_0$$

describe the interactions between the units: for example, these relations may be concerned with the balance of material flows to and from the units. A block angular structure may also arise in a multi-period model in which the x_k represent the decisions to be taken in the kth period.

We will assume in what follows that, for each k, the set $\{x_k : B_k x_k = b_k, x_k \geq 0\}$ is bounded. An easy extension is possible in the case when this is not so; the full details are in Dantzig [1].

5.2.1 The Complete Master Problem

The block angular linear program has

$$(m + \sum_{k=1}^{\ell} m_k)$$

constraints and

$$(\sum_{k=0}^{\ell} n_k)$$

variables. In this section, we show that the problem may be rewritten with fewer constraints but many more variables. This rewritten form is called the Complete Master Problem. It will be shown in the next section that not all the variables of the Complete Master Problem need be considered at the one time and hence that the increase in number of columns is not significant.

Consider the set T_k defined, for each k, by

$$T_k = \{x_k : B_k x_k = b_k, x_k \geq 0\}$$

Then T_k is some convex polytope in n_k-dimensional space. We assume that T_k is bounded. It is well known that each vector in T_k may be represented as some convex combination of the vertices of T_k. That is, suppose we let

$$y_{k1}, y_{k2}, \ldots, y_{kr_k}$$

be all the vertices of T_k: in fact, these are just all the basic, feasible solutions of the constraints $B_k x = b_k$, $x \geq 0$. We may write each x_k in T_k as

$$x_k = \sum_i \lambda_{ki} y_{ki}$$

for some non-negative scalars $\lambda_{k1}, \lambda_{k2}, \ldots$ satisfying

$$\sum_i \lambda_{ki} = 1 \quad .$$

If we were to generate all the vertices of each T_k, we could write the original problem in the variables x_0 and $\lambda_{k_1}, \lambda_{k_2}, \ldots \lambda_{k_{r_k}}$, for each k. The objective becomes

$$\text{maximize} \quad p = c_0^T x_0 + c_1^T y_{11} \lambda_{11} + \ldots + c_k^T y_{ki} \lambda_{ki} + \ldots$$

and the first m constraints become

$$A_0 x_0 + \sum_k \sum_i (A_k y_{ki}) \lambda_{ki} = b_0,$$

but each set of constraints $B_k x_k = b_k$ is replaced by the single constraint

$$\sum_i \lambda_{ki} = 1$$

This rewritten form of the problem has only $(m+\ell)$ constraints but, in general, contains many more variables than the original form. It is called the Dantzig-Wolfe Complete Master Problem. Solving this linear program is equivalent to solving the original problem, since the former is but a rewritten form of the latter. If we knew which variables λ_{ki} were in the optimal basic set for the Complete Master Problem, we would need to generate only those corresponding vertices y_{ki}. It is not possible to achieve this directly but it is possible to restrict attention to a small subset of the variables in the Complete Master Problem. We show this in the next section.

5.2.2 Column generation

Consider using the revised simplex method to solve the complete Master Problem. Only in step 1 (pricing out the columns) is it necessary to have available the columns of coeffients of the non-basic variables. In that step, the algorithm finds the minimum reduced cost d_s, corresponding to the non-basic index s.

In the Complete Master Problem, suppose we partition the multiplier vector π^T into

$$\pi^T = [\pi_0^T, \pi_1, \pi_2, \ldots, \pi_\ell],$$

where π_0 is a vector of length m and π_k is a scalar, $k = 1, \ldots, \ell$. Then the reduced cost, d_{ki}, on the variable λ_{ki} is just

$$d_{ki} = \pi_0^T A_k y_{ki} + \pi_k - c_k^T y_{ki}$$

$$= (\pi_0^T A_k - c_k^T) y_{ki} + \pi_k.$$

Recall that y_{ki} is just a basic feasible solution of the constraints $B_k x = b_k, x \geq 0$. Hence, for a fixed k, we may find

$$\min_i \{d_{ki}\}$$

by minimizing, in x_k, the function

$$(\pi_0^T A_k - c_k^T) x_k + \pi_k$$

subject to

$$B_k x_k = b_k, x_k \geq 0.$$

The optimal solution to this linear program will be at one of the vertices of the feasible region, that is at some y_{ki}. Only if the minimal value is negative need we consider the corresponding vertex in the Complete Master Problem.

We call subproblem k the linear program:

maximize $\quad p_k = (c_k^T - \pi_o^T A_k) x_k - \pi_k$

subject to $\quad B_k x_k = b_k$

$\quad\quad\quad\quad\quad x_k \geq 0$.

For a given multiplier vector π^T, we solve subproblem k and if it has a strictly positive optimal objective value p_k^*, we generate the corresponding column of the Complete Master Problem. If x_k^* is the optimal solution of subproblem k, then the corresponding column of the Master Problem has the vector $A_k x_k^*$ in the first m rows and zeros in the last ℓ rows except for a 1 in the kth position. Thus in each execution of step 1 of the revised simplex algorithm for the Complete Master Problem, only at most ℓ of the non-basic variables need be considered in choosing the minimum reduced cost. This suggests the algorithm of the next section.

5.3 THE DECOMPOSITION ALGORITHM

<u>Step 1.</u> Start with enough columns to form a basic solution to the Complete Master Problem. This is called the restricted master problem.

<u>Step 2.</u> Solve the restricted master problem, which generates a multiplier vector π^T at the end of the solution procedure.

<u>Step 3.</u> Solve each subproblem using the multiplier vector of step 2 to generate the appropriate objective function.

<u>Step 4.</u> If any of the subproblems have (strictly) positive optimal objective values, generate the corresponding column, introduce it into the restricted master problem and go to step 2.

<u>Step 5.</u> If all the subproblems have non-positive optimal objective values, then the current solution is optimal. Recover this optimal solution in its original form from

$$x_k^* = \sum_i \lambda_{ki}^* y_{ki} \quad , \quad k = 1, \ldots, \ell.$$

This decomposition algorithm is just the revised simplex method modified to take account of the special nature of the block angular form of the original problem. Only an outline has been given here. For computational efficiency, some modifications would be made in practice. For example, it is not necessary to solve completely each subproblem in step 3. Any solution which has positive objective value in a sub-problem may be added to the restricted master problem, as in step 4. Indeed, it is often better to add several columns from each subproblem in each iteration of steps 3 and 4. Columns may also be dropped from the restricted master problem if they have not appeared in an optimal basis for several iterations. Because of the large number of columns which may be generated during the algorithm, it may not be possible or efficient to keep a record of the solution corresponding to each column, as required by the given form of step 5. In this case, the solution in the original variables may be recovered by solving each subproblem together with a suitably modified form of the common constraints.

5.4 DECOMPOSITION IN PRACTICE

The Dantzig-Wolfe decomposition algorithm is the most widely used decomposition method but has met with only limited computational success. It is often the case that it is more efficient to solve the original large problem than to use decomposition. This may be due to the present fine tuning of commercially available linear progamming computer codes; similar development of software for decomposition algorithms may lead to increased efficiency of these methods. Williams [4] suggests that decomposition should be attempted only on an experimental basis, at present. If a model is to be used very frequently, such experimentation may prove worthwhile. It is sometimes the case that aspects of structure other than block angularity can be exploited to advantage in solving a particular linear program.

REFERENCES

[1] G. B. Dantzig (1963), Linear programming and extensions, Princeton University Press, Princeton.

[2] L. S. Lasdon (1970), Optimization theory for large systems, Macmillan, New York.

[3] S. S. Rao (1978), Optimization: theory and applications, Wiley Eastern, New Delhi.

[4] H. P. Williams (1978), Model Building in Mathematical Programming, Wiley-Interscience, Chichester.

Chapter 6
Real-time process optimization
D. Foster

6.1 INTRODUCTION

Closed-loop control of continuous industrial processes by a dedicated computer has proved on many occasions to be a highly profitable venture, much of the financial return following from fast, smooth response to process disturbances. When a disturbance occurs, the controlled variables of the process are ramped steadily towards target values supplied to the computer by some external agent, who estimates these target values as the 'best' operating conditions in the new situation. The natural extension of such an application is the provision of these values by the computer itself. For small processes, this may not be a worthwhile undertaking, but for large, complex plants where uncontrolled disturbances are frequent and there are many controlled variables, real-time computer optimization becomes a viable proposition.

This chapter is divided into two parts. In the first section, we discuss the application of an algorithm for steady-state optimization to the real-time optimization of an olefine plant. This is a venture of proven financial return, and one which I.C.I. automatically implements on each new plant of this type. Then a case is presented where the use of dynamic programming to solve an optimization problem connected with the operation of a small industrial power station led to the formulation of an on-line technique for minimizing energy costs.

6.2 REAL-TIME OPTIMIZATION OF AN OLEFINE PLANT

A modern olefine plant, manufacturing olefines by hydrocarbon pyrolysis, presents a difficult problem to process operators who are aiming to achieve the last few percent of profit from the plant. For many processes, large, single-stream units are the order of the day, but for the case in point here, furnace design and operating considerations dictate that multi-furnace configurations have to be used at the reaction stage of the process. For each furnace, there are three independent variables to be controlled on each of several furnaces on a large olefine plant. In addition furnaces are subject to uncontrolled variables associated with changes in feedstock quality and the build-up of carbon deposits in the furnace coils and quench systems. Solving the problem of maximizing the profitability is further complicated by the large number of process limitations imposed by the compressors' tube-skin temperatures, furnace firing and quench systems, coking rates and so on, which are themselves changing in time because of variations in cooling water temperature, carbon lay-down and feedstock quality.

The problem of how to get maximum instantaneous profit out of a large cracker can be solved by building a closed-loop computer into the furnace control system. The computer optimizes furnace operation for continuous maximum profitability within the process limitations, taking into account changes in product realizations and prevailing market situations. Many operators of petrochemical processes have constructed mathematical models of their plants for production planning purposes which enable them to define the optimal monthly strategy of plant operation. The justification for real-time plant optimization by a dedicated computer lies in the increased profitability which is obtained by close attention to plant operating details which only such a system can provide. The intimate links between plant and computer enable the large amount of data which is necessary to define optimal operation to be collected relentlessly. Once defined, the relatively simple step of outputting directly to the process controllers ensures that these optimal conditions are rigorously imposed on the plant.

It is important to realize that such a system does not operate in isolation. It is the lowest level of a hierarchy of planning models, and it communicates with its neighbours through sets of product values and break points which reflect the environment in which the plant exists. For example, can we sell ethylene? If not, are we able to store it or must we burn it? Depending on the answers to such questions, the worth of major components may vary between chemical and fuel value. For each product of each plant, the effect on overall gross margin of unit increase in the flow of the product is called its marginal value, and is dependent not only market conditions but also on active constraints within the plant. In general, marginal values apply only to a limited disturbance about the set of operating conditions at which they are generated, but they provide a more realistic set of values with which to optimize than the chemical values of the various products. The marginal values and costs used in real-time optimization applications are generated on a regular basis by the business-area LP model, which includes representations of the plants in question.

6.2.1 An olefine plant at Wilton

Fig 6.1 shows a simplified flow diagram for an olefine plant. Parallel naphtha and ethane cracking furnaces feed products through the primary fractionator to a multi-stage process gas compressor, and then into the gas-separation phase of the process. In each furnace, there are tubular reactors through which naphtha (a light cut of crude oil) and steam are passed and heated to a high temperature. The coils join in pairs before leaving each furnace, and the two resulting process streams are fed to quench boilers. The quench boilers cool the process streams by heat exchange against saturated water. This cooling stops the reaction and recovers energy in the form of high-pressure steam for turbine drives and pumps. The combined process stream is further cooled by the primary fractionator, where heavy fuel oil is condensed out. The heat removed by this tower can be a limitation on plant throughput, especially during the summer months when heat exchanger capability is reduced. The process stream is then sequentially compressed by a turbine-driven compressor. Separation of the final products is achieved by cooling and distillation.

Historically, the scope of any optimization has always been limited to the 'hot end' of the plant, that is to the selection of front-end independent variables. Although there are several degrees of freedom in the downstream section of the plant (distribution of cooling capacity, compressor speed, etc.) values for these quantities are chosen according to some specific rules. The major contribution of the downstream units to the optimization is through any process limitations they may provide. Examples are the process gas compressor throughput limit, the primary fractionator capacity and the ethylene rate from C_2 splitter.

6.2.2 The computer control scheme

The computer control scheme uses a mathematical model of the process to evaluate plant profit and constraint proximity for any set of values of the process variables. By means of an

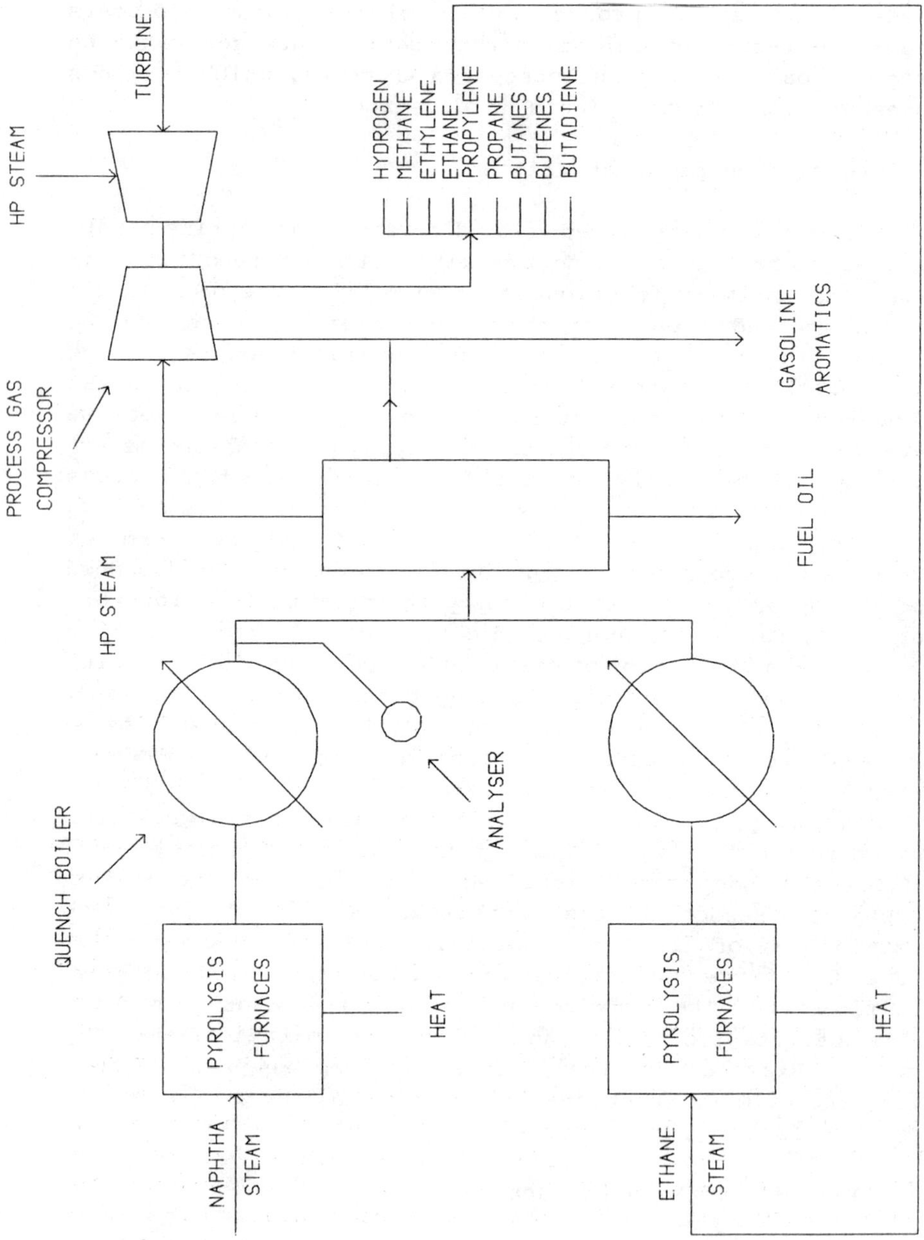

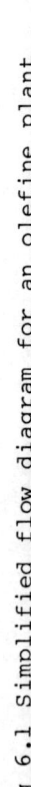

Fig 6.1 Simplified flow diagram for an olefine plant

optimization algorithm, the computer searches for those values of the controlled variables which achieve maximum profitability. Set point adjustments are then made automatically on the process controllers to hold the plant at maximum profit within the process constraints. As plant performance changes because of coking or feedstock variation the model is adapted to reflect this uncontrolled disturbance. Well-proven analysers are essential, and relay data on the cracking patterns of individual furnaces to the computer. Certain key coefficients within the model are adjusted so that they continuously represent plant performance.

So far, the discussion has been of what is apparently a steady-state optimization problem in the commonly accepted sense, but in a real-time implementation the ultimate goal of solution is very rarely achieved. In practice, the number of independent variables of the problem may change as furnace variables are removed from computer control (and hence from the optimization) for control-loop maintenance. The extremal process constraints are subject to discrete changes by operator intervention. The reason for this is that in spite of the model updating done by the computer, there will remain some inconsistency between model prediction and plant performance. Thus as the optimized variables are changing in order to travel along a constraint (as modelled), the plant itself may begin to stray from actual limitation, and an adjustment in level will be necessary to ensure that what is limiting in the plant is also limiting in the model. For on-line implementation, the optimizer must be designed to accept intermittent changes of this nature.

It is imperative that the optimizer be capable of absorbing these continual interruptions as quickly as possible, for the reason that set-points associated with a previous problem (besides being sub-optimal) may be grossly infeasible. The output of infeasible operating modes is disallowed. Normally the optimization algorithm would be structured to allow a response time of the order of minutes. This gives sufficient time to make worthwhile progress in optimizing, but does not cause a long delay when responding to a change in problem definition. In assessing the length of this delay, consideration has been given to the importance of the relationship between plant personnel and the computer. An on-line system which is to earn the respect of its users must be seen to be alive and active, and not, for example, spend thirty minutes solving a problem which has become obsolete because a furnace has been taken off-line. To achieve an acceptable maximum response time, the optimization may be conceived as broken down into isolated bursts of computation. At the end of each burst, plant conditions are investigated automatically by the computer, and if there are changes which affect the optimization significantly, a new problem is formulated by the computer. It is at this point that it may be necessary to reset certain internal parameters of the optimization to partially re-start the search.

The chosen optimization algorithm must, of course, be able to handle inequality constraints. Explicit and implicit inequality constraints are treated in different ways. As we have seen in chapter 1, almost all algorithms involve a univariate search in some chosen direction as part of each iteration. During this search, if an upper bound is encountered, the limit is adopted. Only if the sign of the step direction component at some future iteration indicates that it is worthwhile to leave the limit is the variable allowed to move. Evidently, if the explicit bounds on independent variables were commonly active constraints, it is conceivable that this procedure would lead to oscillation, with bounds alternately joining and leaving the constraint set. On the other hand, this method of handling the explicit constraints cuts down the number of constraints in the calculation, and it is worth accepting here for this reason, because unless a plant is incorrectly designed, furnace limitations should not be explicit. Furnace throughput should be restricted by firing heat, coking, etc., which are the implicit constraints. Implicit constraints have always been handled in alternative ways. The maximization algorithms have always been chosen to be as simple and compact as possible. With many variables and perhaps twice as many constraints, there was no space for the storage of large matrices or sets of directions spanning the optimization space. It is recognized that with the advent of cheap computer store, this may be no longer true. Even so, the value of updated matrices will be lost if at the end of an optimization burst it is discovered that there has been some change in problem definition. Hence methods that do not involve matrix manipulation, and use modest amounts of computer storage are preferred.

It is worthy of note that great speed in on-line optimization is not essential. The rate at which plant variables may be adjusted through their controllers is very limited. Provided optimization progress is sufficient to lead the plant, this is satisfactory. The procedure of on-line optimization may be viewed as one of pursuing a changing objective, and one that will rarely be reached.

The importance of operation under the correct constraint cannot be over-emphasized. At best, the computer may simply choose a slightly sub-optimal strategy. At worst, operator interference can cause a gradual divergence between recommended conditions and optimal conditions. Under these circumstances, the optimizer could become a liability, forcing the adoption of operating policies which would never have been considered in the absence of the computer.

6.3 OPTIMAL OPERATION OF A SMALL INDUSTRIAL POWER STATION

All of the existing on-line optimization applications in Petrochemicals Division have involved the tailoring of an algorithm for non-linear optimization to the specific needs of on-line implementation. Recently, however, investigations into ways of making more efficient the production of energy to meet the needs of an industrial site have revealed that dynamic programming may have a useful part to play. The remainder of this chapter is devoted to an example which illustrates the use of Bellman's method for the solution of stagewise dynamic optimization problems. It is hoped to show how experience with this approach led to the development of a relatively simple solution technique which could form the basis of an on-line energy optimization scheme.

The I.C.I. power station in question is depicted schematically in Fig 6.2, and supplies three grades of steam (high, intermediate and low pressure h.p., i.p., and l.p.) to an industrial site, together with a certain amount of generated electricity. The steam requirements are satisfied by generating sufficient high-pressure steam in three oil-fired boilers. This is let down through two turbo-alternator sets operating in series on the steam side, to provide the intermediate and low-pressure steam required by the site. The turbines are back-pressure controlled, and so the steam demands determine the electrical power generated by the sets. The i.p. and l.p. steam required by the process plant at any given time fix a base electrical power generation, which can be reduced only by by-passing either or both turbines. These steam demands are dependent on the production rates of the process plants, and may be estimated in advance for a given production plan. High production levels imply high steam demands. When the steam demand is high, it is possible, though not desirable, that this base level of electrical generation may exceed the site requirements for electrical power. In that case, the surplus

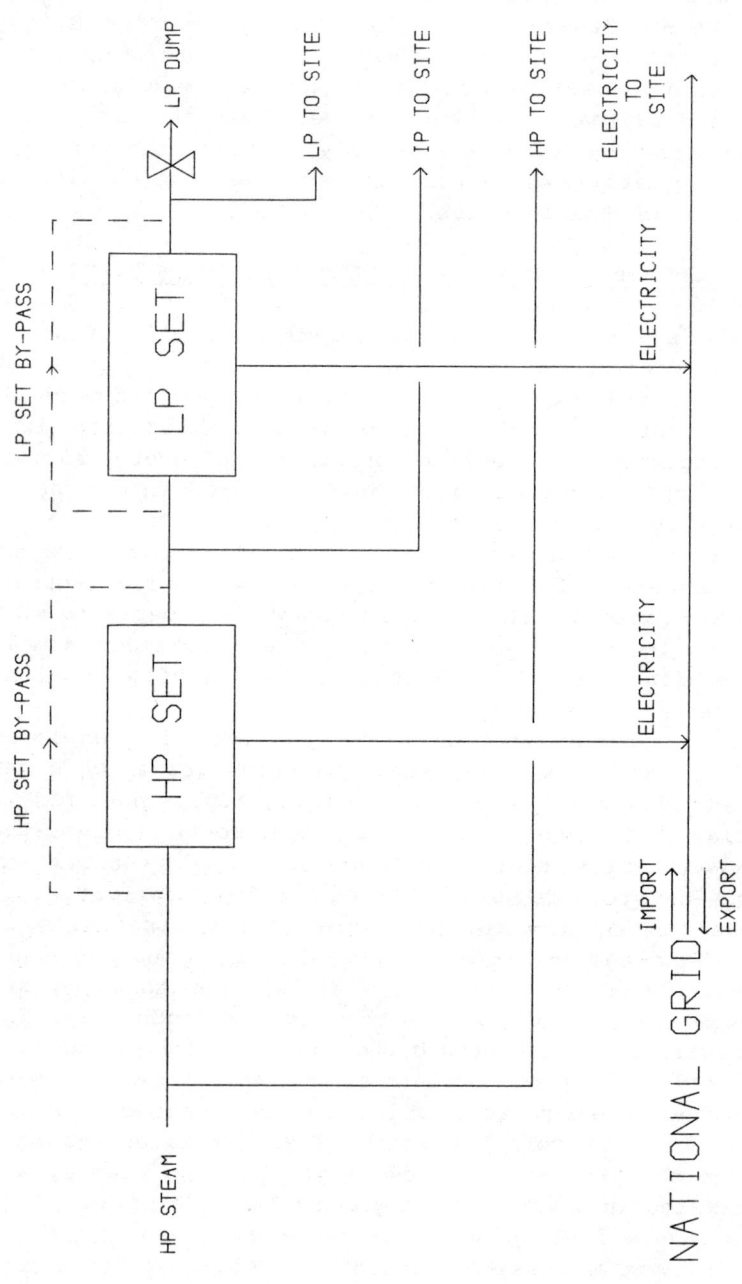

Fig 6.2 Schematic of small industrial power station

power is exported to the C.E.G.B. for a small revenue. Otherwise, the base quantity of electricity will be below requirements, and the deficit must be made up either by import from the National Grid, or by additional generation. In the base case the l.p. steam valve is closed. When the valve is opened and l.p. steam is exhausted to the atmosphere, the boiler pressure control loop will cause more h.p. steam to be generated by the boilers. The flow through each set will increase until at least one set reaches its power limit, when additional dumping will result in the operation of the by-pass round the limiting set, and no more power will be available from that generator. Dumping beyond this point may be worthwhile only until one of three events occurs:

(a) The power limit on the other set is reached

(b) The l.p. dump valve can pass no more flow.

(c) The boilers can generate no more steam.

If one of these three conditions is reached and the site electrical demand is still not met, import from the Grid is inevitable.

The choice of the extent of l.p. steam dumping (and hence of import of electricity) to meet a given deficit should be made on a financial basis. If it were the case that the cost of imported electrical power were constant, then the problem of minimizing the cost of supplying electrical energy over a given time period would simply be one of deciding the best level of l.p. steam dumping at each instant. Each choice would be a choice in isolation, and would not need to take into account performance during the rest of the time period. Unfortunately, the "Maximum Demand Tariff" monthly charging scheme operated by the C.E.G.B. complicates the choice, and the optimal policy at any instant may depend on the performance during the rest of the month, and may not be entirely obvious.

6.3.1 The optimization problem

Provided with a demand profile for site utilization of steam and electricity over a given period (one month), the purpose of the optimization is to select a profile of l.p. steam dumping, and hence of levels of electrical power generation, so as to minimize the cost of supplying the required amount of energy over that period. The minimization of this overall monthly cost is an essentially dynamic optimization. This is so because the C.E.G.B. cost is computed from functions of time which are known only when the import pattern for the whole period is known.

We divide the period into n equal intervals, of length Δ. Typically, Δ may be the length of a shift and n about 90 shifts per month. Suppose that the demands in interval i are SI_i, SL_i and D_i of i.p. steam, l.p. steam and electricity respectively, for $i = 1, 2, \ldots, n$. It is desired to determine the levels of combined generator output $\{G_i : i = 1, 2, \ldots, n\}$ which minimize the overall cost of meeting the demands throughout the month.

The levels of electrical import in each interval are denoted by $\{I_i : i = 1, 2, \ldots, n\}$, and

$$I_i = \max\{D_i - G_i, 0\},$$

while the exports $\{E_i : i = 1, 2, \ldots, n\}$ are given by

$$E_i = G_i + I_i - D_i.$$

The two parameters which determine the C.E.G.B. monthly charge are "Maximum Demand", M and "Total Units Import", I, given by

$$M = \max\{M_0, I_1, I_2, \ldots, I_n\},$$

$$I = \Delta(I_0 + \sum_{i=1}^{n} I_i),$$

where M_0 and I_0 are "already incurred" values of these quantities. M is in reality more complicated than this, being based on twice the number of units imported during any half-hour period, but for the purposes of this analysis its definition as peak import will suffice.

The revenue from electricity exported is

$$r\Delta(E_0 + \sum_{i=1}^{n} E_i)$$

where r is the revenue per MWH. We write

$$E = E_0 + \sum_{i=1}^{n} E_i.$$

Thus for a trial generation profile $G_1, G_2, \ldots, G_n$ the overall cost is

$$f(G_1, G_2, \ldots, G_n) = \Delta(\sum_{i=1}^{n} g_i(G_i) - rE) + \phi(M, I).$$

Here $\phi(M,I)$ represents the as-yet undefined C.E.G.B. monthly cost for I units of import at a maximum demand M, and $g_i(G_i)$ represents the the cost of generating electricity at a level G_i during interval i. The function g_i is dependent on steam demands during interval i and so is unique to that interval. We now discuss the form of ϕ and the functions $\{g_i\}$ in more detail.

The function ϕ: The C.E.G.B. charges for imported electricity on the basis of a monthly costing scheme known as the Maximum Demand Tariff, illustrated in Fig 6.3. At the end of the month, a histogram profile of the import during the month determines M and I. The cost is divided into two parts. There is a maximum demand charge, which is a piecewise linear function of M. Over each section of the curve, the cost per MW for increasing maximum demand is a constant. The second part of the monthly cost is the units charge. Again, this is a piecewise linear function of I. The units charge is further complicated by the fact that the length of each plateau of constant import cost per unit is related to the maximum demand M. Typically, the first 200M MW are charged at highest rate, the next 200M MW are charged at a lower rate, and so on. Thus, in considering whether or not it is worthwhile to increase maximum demand, it must be remembered that the charge for units already used may increase. Further, if it is possible to operate at low maximum demand, yet achieve a total import which invokes the least marginal cost of import (that is, achieves the third cost plateau), then marginal imported power is at its cheapest.

The functions $\{g_i\}$: Fig 6.4 illustrates the dependence of generation costs on generator output. The site steam demands in interval i establish a base level of electricity generation L_i. At this point, the l.p. dump valve is closed. As the valve is opened and l.p. steam is dumped to the atmosphere, the electrical output from each set will increase until at least one set achieves its maximum power output. Further dumping will cause the by-pass around that set to open, and the power output of that generator will remain at its maximum. Call this level of combined output H_i. At this point, the marginal cost of

116 Real-time process optimization

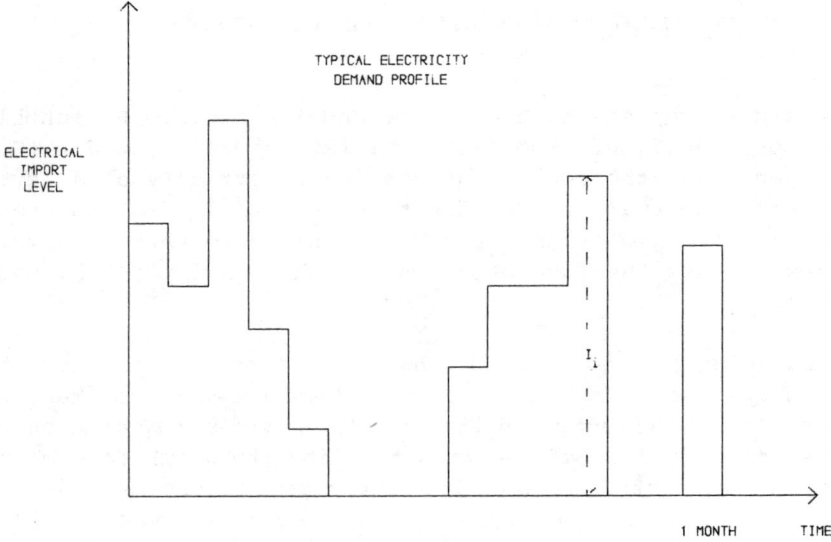

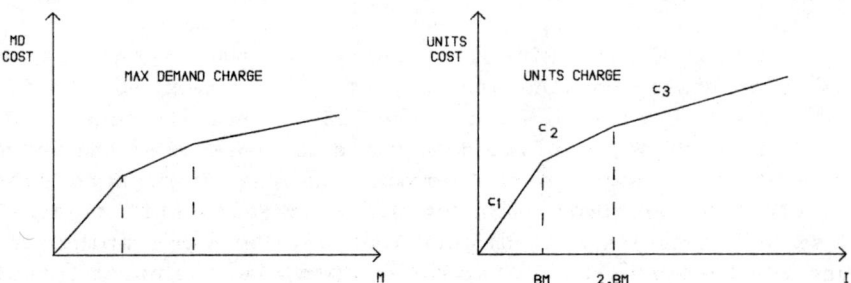

Fig 6.3 The Maximum Demand tariff.

generation increases discontinuously. Dumping beyond this point will be possible until the maximum output of the remaining set is reached or until the dump valve can pass no more flow. Let

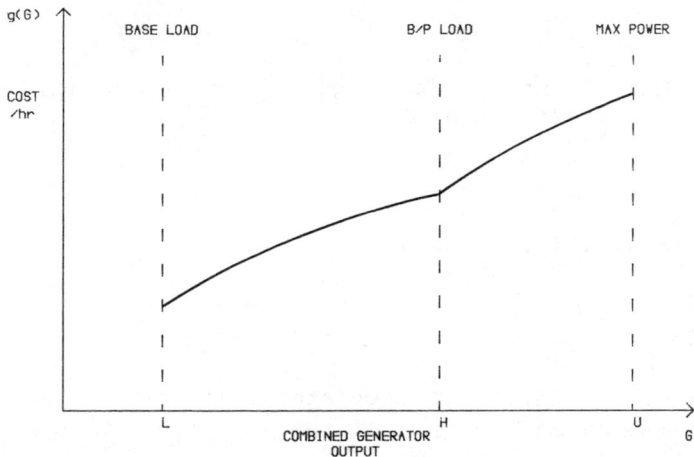

Fig 6.4 The cost of generation.

this maximum level of generation be denoted by U_i. It is easy to show that if the functions h(P) and l(P) represent the steam consumptions of the sets when the power available from an individual set is P MW, then g_i is defined according to one of three conditions. Evidently, when selecting the optimal levels of generation $\{G_i\}$, they must be restricted so that

$$L_i \leqslant G_i \leqslant U_i$$

An important observation is that the differentiable sections of g_i are concave, and if we are searching for a minimum of a function of this type, it must be at either L_i, H_i, or U_i.

Having defined the cost function in terms of the controls $\{G_i\}$, we note that its minimization could be regarded as a problem in static optimization. The chosen method would, however, need to handle the multitude of discontinuities of derivative that may exist at the solution. Methods where analytical niceties are assumed do not seem suitable, while methods designed specifically for non-differentiable optimization have not so far proved very efficient, as was remarked in chapter 1.5.3. It is perhaps appropriate here to say that most industrial optimization problems cannot be formulated to satisfy nice analytic conditions. Step changes in unit prices and costs, for example, are common.

For the current problem, Bellman's dynamic programming offers a solution. The overall minimum cost for n stages, with initial values of α for Maximum Demand and β for total import, is defined as

$$f_n(\alpha,\beta) = \min_{G_1,G_2,\ldots,G_n} \left\{ \Delta \sum_{j=1}^{n} [g_j(G_j) - rE_j] + \phi(\alpha_n, \beta_n) \right\}$$

where

$$\alpha_n = \max\{\alpha, I_1, I_2, \ldots, I_n\},$$

$$\beta_n = \beta + \sum_{j=1}^{n} I_j.$$

For brevity, the factor Δ associated with the total import argument is subsumed in ϕ. By the Principle of Optimality (chapter 2.3.2),

$$f_n(\alpha,\beta) = \min_{G_n} \{\Delta[g_n(G_n) - rE_n] + f_{n-1}(\max\{\alpha, I_n\}, \beta + I_n)\},$$

where $f_{n-1}(\alpha', \beta')$ is the optimal cost of meeting the demands in periods $1, 2, \ldots, n-1$ with initial values of α' and β' for Maximum Demand and total import. More generally, if $f_k(\alpha', \beta')$ is the optimal cost of meeting the demands in periods $1, 2, \ldots, k$ with initial values α' and β' for M and I, then

$$f_k(\alpha',\beta') = \min_{G_k} \{\Delta[g_k(G_k) - rE_k] + f_{k-1}(\max\{\alpha', I_k\}, \beta' + I_k)\}.$$

Given values for f_{k-1}, f_k can be derived by considering all possible values of $G_k \in [L_k, U_k]$, and picking the best. The value of f_n is required for only one pair of arguments, I_0, M_0. Clearly, as k increases, the number of values of f_k required increases. The largest number of evaluations is needed for f_0, which is identical to ϕ, the C.E.G.B. cost function, and is the cost of meeting demand for zero stages with already incurred values α', β' for M and I.

It is a simple matter to derive the ranges of arguments for which the values of the functions $\{f_k\}$ are needed. In fact, $f_k(\alpha', \beta')$ is required for

$$\sum_{j=k+1}^{n} \max\{M_0, D_j-U_j\} \leq \alpha' \leq \sum_{j=k+1}^{n} \max\{M_0, D_j-L_j\},$$

$$I_0 + \sum_{j=k+1}^{n} \max\{0, D_j-U_j\} \leq \alpha' \leq I_0 + \sum_{j=k+1}^{n} \max\{0, D_j-L_j\}.$$

Hence the minimization of the overall cost function can be done by a three-step algorithm.

1. Determine $f_0 \equiv \phi$ for an appropriate range of arguments, using a suitable discretization interval.

2. Use the dynamic programming recursion to find required values of $f_1, f_2, \ldots, f_n$. For each pair of argument values, store the minimizing G_k. Clearly the determination of f_k requires that f_k and f_{k-1} be in core together.

3. Having found $f_n(M_0, I_0)$, reprocess tables to find $G_n, G_{n-1}, \ldots, G_i$. There may be several choices for each G_i.

The two important parameters emerging from this solution are maximum demand and marginal import cost. The marginal import cost, denoted by c, identifies the optimal import cost plateau for the anticipated production pattern. Clearly, the real production pattern will be somewhat different, but it is assumed that all major events (like plant shutdowns) have been incorporated. This means that all important mal-distributions in the steam demand have been included. In other words, the possibilities for any forced maximum demand have been accounted for. In order to develop an on-line system, one makes the assumption that for small changes in the anticipated production pattern, the same optimal maximum demand and marginal import cost hold.

Fig 6.5 illustrates the proposed on-line system, using the two key parameters derived by running the DP algorithm off-line, with the anticipated steam demands. If any major deviations from the plan occurred during the month, this would be repeated and new values used for the remainder of the cost period. A typical major upset is generator outage which forces a high level of import, albeit for only thirty minutes.

With the current targets for the key parameters, the on-line system needs continuously to minimize the overall cost for small deviations from plan. However, since maximum demand is now effectively fixed, the minimization may be decoupled into a sequence of independent problems of the type

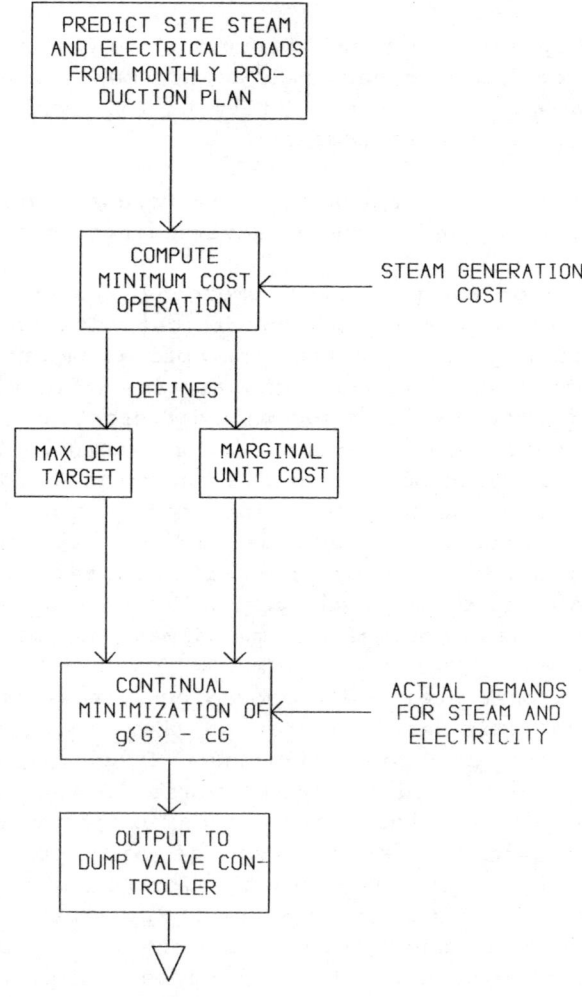

Fig 6.5 The on-line system.

Minimize $g(G) + (c-r)\max\{D-G, 0\} - rG$

subject to $G \leq U$,

$G \geq L$.

In practice, generating to export is never profitable, and so this simplifies further to

Minimize $g(G) - cG$

subject to $G \leq U$,

$G \geq L$.

because $G \leq D$. This requires just a direct comparison of generating cost and marginal import cost over the allowable range of generating levels.

Since g is concave, so is $g(G) - cG$. Hence the minimum sought must be at one of the three points $G = L$, H, or U. There are three possibilities that may arise, depending on the position of the by-pass open point H relative to the lower and upper levels L and U. In Fig 6.6, each curve on each diagram is a possible form of $g(G) - cG$ over the interval (L, U). Which one applies evidently depends on the import cost c and the fuel oil value implicit in g. The important point is that with the prescribed values of maximum demand and marginal import cost, the on-line minimization is trivial.

In the final system, which is as shown in Fig 6.5, a small computing device accepting as inputs measured values of the steam and electricity demand continuously minimizes $g(G) - cG$. It uses a target value for maximum demand and a marginal import cost derived by solving the anticipated problem on a larger external computer. At regular intervals, it outputs the optimal l.p. dump valve set point to the valve controller to set the optimal level of electrical generation. The set point may correspond to zero valve opening. Industrial power stations, in their efforts to minimize energy supply costs, may find themselves in the anomolous position of having to throw away energy to achieve their objectives!

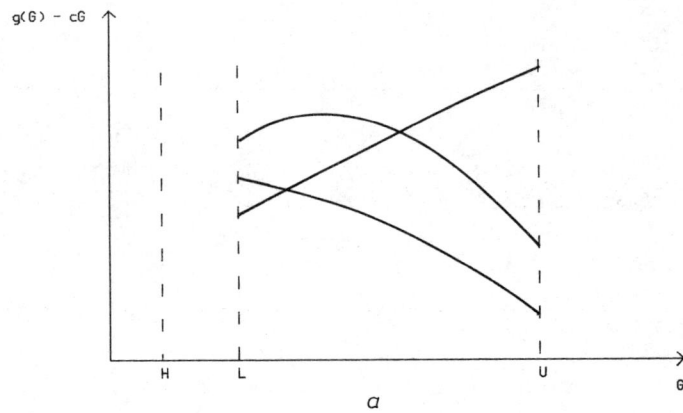
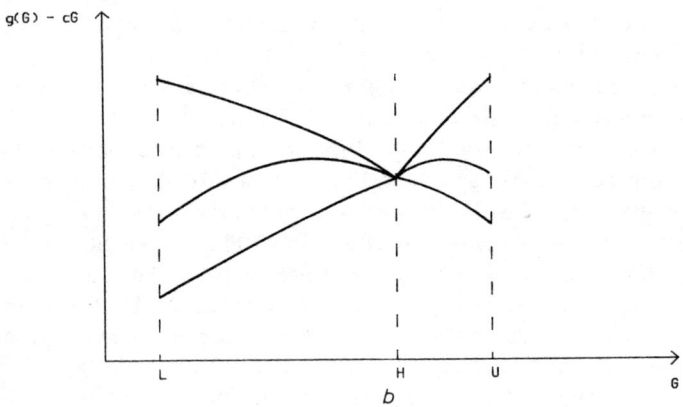
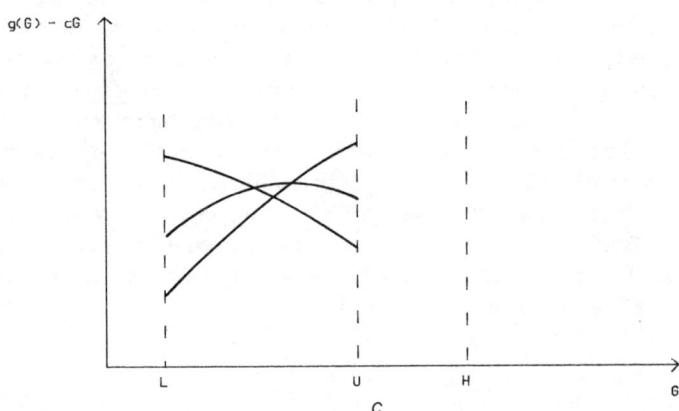

Fig 6.6 Forms of g(G)-cG

Chapter 7
Control theory in macroeconomics
M.B. Zarrop

7.1 INTRODUCTION

The entity known as the National Economy is arguably the largest and most complex system over which man has attempted to exert some degree of influence. Over the past thirty years, and particularly since the mid 1960'S, there has been an increasing effort by both engineers and economists to use the tools provided by optimization and control theory in the field of economics, both at the level of the individual firm (microeconomics) and the level of national policy formulation (macroeconomics).

A recent survey by Derakhshan [8] shows that over 1600 papers, books, dissertations, etc., have been produced in this interdisciplinary area, mainly since 1966, with a rapid expansion in the last decade. Undoubtedly, many of those working in the control field considered it insufferable that the power of modern computers and modern control theory, capable of guiding men to the moon, was not being used to assist in controlling unemployment and inflation, particular as the economists appeared unable to provide an effective antidote to these symptoms of an increasingly worsening world economic situation.

On the other hand, economists have been justifiably suspicious of the glib and easy solutions sometimes proposed by the systems men and also of some of the more sophisticated schemes advocated. Even more seriously, harsh economic reality has, in the view of some economists, not only placed a question mark over the usefulness of macroeconomic models for forecasting purposes but has also shown the futility of discretionary policy, i.e. Keynesianism and government intervention.

Nevertheless, it is safe to say that macroeconomic models are here to stay and, despite the agonising of politicians and economists, decisions on economic policy will be taken which are based partly on computer manipulations of these models.

It is with the quality of these decisions that we are concerned.

7.2 MACROECONOMIC MODELS

The construction of an econometric model necessarily starts with the collection and transformation of a large amount of economic data. As in every discipline, the model builder is guided by theory in selecting those data streams necessary for his task. The U.K. system of national accounts follows broadly the principles recommended for international use by the United Nations and the O.E.C.D. Within this framework relationships between different sets of variables are suggested by economic theory and can be tested using statistical procedures.

Macroeconomic variables fall into three categories: endogenous variables (targets, outputs), instruments (control inputs) and purely exogeneous variables (other inputs). This division is not absolute. Even within a single model, variables considered endogenous over some period of time (e.g. a 'floating' exchange rate or wages) may subsequently be exogenised (fixed exchange rate, incomes policy) and vice versa. Purely exogenous variables include factors outside the control of the national government, such as world prices, and also dummy variables introduced to correct for the distorting effects of sudden shocks, such as strikes.

Model equations are of two types: identities and behavioural relationships. Identities are exact, non-estimated relationships such as

Personal disposable income

$$= \text{(gross personal income)} - \text{(taxes)}, \quad (7.1)$$

while the other equations give expression to what economic theory says about the behaviour of economic agents and may involve the estimation of parameters from data. For example,

$$\text{Investment} = \alpha + \beta \times \text{(interest rate)}. \qquad (7.2)$$

(Note that we expect β to be negative.) Thus the model builder constructs from discrete data a structural model of the form

$$Y_1 = f_1(Y,U,E,\theta) + \text{error} , \quad (7.3a)$$

$$Y_2 = f_2(Y,U,E) , \quad (7.3b)$$

where $Y = (Y_1^T, Y_2^T)^T$ is the endogenous vector, partitioned between the behavioural equations and identities, and U, E, θ are the vectors of instruments, purely exogeneous variables and parameters respectively. Equations (7.3) include dynamic models if the vectors are stacked up over a number of periods, as

$$Y = (Y(1)^T, Y(2)^T, \ldots, Y(N)^T)^T \in \mathbb{R}^{rN} , \quad (7.4)$$

where $Y(k)$ denotes the r-vector of endogenous variables for time $k = 1, 2, \ldots, N$.

The size of model used for forecasting, particularly in the industrialised world, tends to be large compared to those models used in control engineering. The table below lists the sizes of the three main U.K. forecasting models.

Model	Behavioural equations	Identities
Treasury	300	283
London Business School (LBS)	87	187
National Institute (NIESR)	58	124

A number of points can be made concerning these models and their possible use for control purposes:

(a) The models are discrete-time models, based on quarterly data that is fairly complete only since 1955. Thus the data stream is short, leading to estimation problems.

(b) Data series may be revised or corrected periodically, leading to the need for re-estimation of some or all of the behavioural equations (observation error).

(c) Models are generally nonlinear in the variables, but expressed in a form that is linear in the parameters and estimated by ordinary least squares methods equation by equation. Nonlinearity is often introduced through log transformations on the data to induce stationarity.

(d) The models are solved by iterative procedures, usually Gauss-Seidel. For example, the current LBS model needs an average of 15 iterations to converge at each time period and this takes about 0.5 CP second on Imperial College's CDC 6500.

(e) The models are improper, in that some exogenous variables appear to have an instantaneous impact on certain targets. This can occur because of accounting conventions or when an exogenous change takes less than one period to make itself felt.

(f) The possibility of active experiment design, for example injecting a test signal to probe the system, is unlikely to be countenanced in the near future. It may be useful, however, to sample certain economic variables more frequently.

(g) The economic process is probably time-varying, whereas models are estimated as time-invariant.

(h) Model forecasts are affected by a number of error sources. Primarily, however good the model, forecasts of exogenous inputs are always open to question (e.g. oil prices!). In addition, modelling errors occur and are reflected in the standard errors of equation residuals and parameter estimates, but are usually ignored when forecasting.

(i) The construction of an objective function for economic policy optimization poses problems. It is unlikely that the policymaker (ultimately, the Chancellor of the Exchequer) could express his wishes in this form, even if he wanted to.

(j) Optimal control assumes a centralised controller – in this case, the government. In reality, we have a number of conscious agents (banks, corporations, TUC, CBI, etc.) each with important economic leverage and operating with differing models and objectives. In particular, it has been suggested that the <u>expectations</u> of economic agents may nullify the ability of a government to use 'optimal' policies at all.

At this juncture, we make only two points in response to the above rather long list of obstacles and criticisms that faces us. Firstly, in comparison to engineering systems, there is a long time between samples in which to complete and assess future policy. Thus, sophisticated schemes, if considered relevant, can be entertained. Secondly, but most important, a relatively small improvement in economic performance can mean a big pay-off.

7.3 CONTROL VIA LINEAR MODELS

The greater part of control theory has been concerned with linear models and the starting point for many economic control exercises both in the U.K. and elsewhere has been the construction of a suitable linear model of manageable dimensions.

In the U.K., Imperial College's Programme of Research into Optimal Policy Evalution (PROPE) began its work ten years ago with the construction of a sequence of linear models leading up to the treatment of policy formulation as an LQG exercise. The model was estimated in rational structural form (Wall, [14]) using Maximum Likelihood estimation on blocks of equations, leading to a model of the form

$$y(k) = A(z^{-1})y(k) + B_1(z^{-1})u(k)$$
$$+ B_2(z^{-1})e(k) + C(z^{-1})\omega(k) , \qquad (7.5)$$

where $y(k)$, $u(k)$, $e(k)$, $\omega(k)$ are the vectors of endogenous variables, instruments, exogeneous variables and stochastic disturbances respectively, A, B_1, B_2, C are rational matrix operators of appropriate dimensions, and z^{-1} is the backward shift operator. Those equations corresponding to identities have a zero row in C.

The determination of structure involves a combination of prior economic information and repeated use of single equation model building techniques (Box & Jenkins, [5]) to find a parsimonious model that gives a satisfactory account of the data yet retains economic validity. Models of the form (7.5) can be cast in a minimal state space form (Preston & Wall, [11])

$$x(k+1) = Fx(k) + G_u u(k) + G_e e(k) + G_\omega \omega(k) , \qquad (7.6a)$$

$$y(k) = Hx(k) + D_u u(k) + D_e e(k) + D_\omega \omega(k) , \qquad (7.6b)$$

The LQG problem is that of selecting $\{u(k), 1 \leq k \leq N\}$ to minimize the cost function J given by

$$J = E\{\frac{1}{2} \sum_{k=1}^{N} [\delta y(k)^T A(k) \delta y(k) + 2\delta u(k)^T C(k) \delta y(k)$$

$$+ \delta u(k)^T B(k) \delta u(k)]\},$$

subject to the constraints (7.6), where δ denotes deviation from the desired trajectory and E denotes the expectation operator. This is easily transformed into the standard state tracking problem of control theory.

Although (7.6) is a stochastic model, its form implies <u>full state information</u>, thus avoiding the use of Kalman filtering techniques. Eliminating ω(k) yields the steady-state Kalman filter form

$$x(k+1) = Ax(k) + Bu(k) + Ce(k) + Ky(k), \quad (7.8)$$

where

$$A = F - KH,$$

$$K = G_\omega D_\omega^{-1},$$

$$B = G_u - KD_u,$$

$$C = G_e - KD_e$$

and the matrix A is asymptotically stable. Assuming an incorrect state estimate at the data origin, the error dies away as (7.8) is run forward to the control origin. Thus the initial state $x(1)$ can be assumed known and $x(k)$ can be calculated exactly from input-output data for $k > 1$ using (7.8).

The dynamic programming solution to this LQG problem is straightforward (Wall & Westcott, [15]) and yields an optimal state feedback solution of the form

$$u^*(k) = L(k)x(k) + h(k), \quad (7.9)$$

where $L(k)$, $h(k)$ are the control and tracking gains respectively at time k, calculated by backwards recursion. Of course, (7.8) and (7.9) can also be used to generate the optimal open-loop control sequence. This framework has been used with linear models with up to 60 states but using no more than five weighted targets and five instruments. A provisional list of weighted variables might include:

Targets: Inflation, unemployment, GDP, balance of payments, money stock, short term interest rate

Instruments: Government expenditure, exchange rate, taxes, minimum lending rate, public sector borrowing requirement.

Money stock is sometimes referred to as an intermediate target because its control (according to the monetarists) has a moderating effect on the inflation rate.

A priori, we can assign only rough constant diagonal weights in the quadratic cost function (7.7) and we may have to adjust these to get acceptable trajectories (see section 7.6). Of course, we can complicate the problem by imposing various inequality constraints, but this destroys the LQG framework and imposes a larger computational burden. A useful device, related to the penalty function approach described in chapter 1.4, which avoids this is to introduce dummy targets. For example, if an instrument fluctuates too wildly we can introduce a target equal to the first difference of the instrument and weight any deviation from a zero trajectory. This device is also useful for constraining the tax rate instrument to change only once a year by heavily weighting its first difference in every period but the budget quarter. Similarly, money supply can be constrained to follow a constant (unspecified) growth rate if its second difference is weighted to track zero in all periods but the first.

7.4 PERTURBATION MODELS

Much effort is expended in constructing economic models (linear or otherwise) that are credible both from the point of view of economic theory and forecasting performance. It can be argued, however, that while the forecaster tends to concentrate on the statistically efficient estimation and prediction of a large number of targets, the control analyst is interested primarily in the robust identification of the mapping between the chosen instruments and the subset of targets to be controlled.

Holly et al [10] propose a perturbation method for generating a suitable control model. Each instrument is perturbed independently by discrete-time white noise over the control period, so that

$$E[\tilde{u}_i(k)\tilde{u}_j(s)] = \lambda_i \delta_{ij} \delta_{ks} , \qquad (7.10)$$

where ~ denotes perturbation and λ_i is chosen so that the perturbations are around 10% (say) of some mean level for each instrument. The orthogonality property (7.10) considerably simplifies the estimation problem. Each transfer function $u_i \rightarrow y_j$ is assumed rational and can be estimated separately. Finally, a perturbation model is built up by superposition and the equation residuals estimated as ARMA (auto-regressive moving-average) models if required. Note that if the model is realised in state space form as in section 7.3, then the initial (perturbation) state is zero <u>without</u> approximation.

7.5 CONTROLLING A LARGE MODEL: I

Given a quadratic cost function and linear model, the LQG framework generates both open-loop and feedback control laws. How well do these perform if they are used to control a large, non-linear forecasting model?

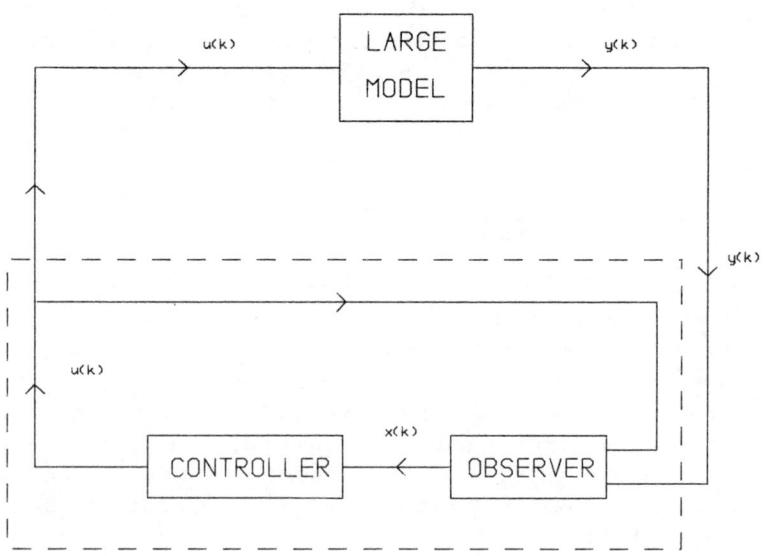

Fig 7.1 Linking of observer to large model

The form of the linear model (7.6) leads to (7.8) and this suggests a simple form of state observer for the large model that can be employed in conjunction with the optimal control law (7.9). The "link-up" is shown schematically in Fig 7.1.

In detail, the following heuristic control algorithm emerges:

<u>Phase 1</u>   (Linear model)

(i)   Set up quadratic cost function and solve LQG problem for N periods;
(ii) Store control and tracking gains and the initial state x(1).

<u>Phase 2</u>   (Large model)

(i)   Set j = 0;
(ii)   Set j = j + 1;
(iii) Set $u^*(j) = L(j)x(j) + h(j)$ and input to large model to generate y(j);
(iv)   If j = N, STOP.
(v)   Set $x(j+1) = Ax(j) + Bu^*(j) + Ce(j) + Ky(j)$;
(vi)   Go to (ii).

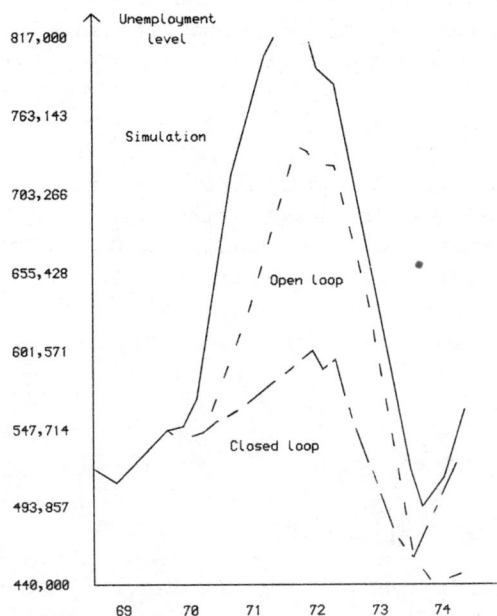

Fig 7.2   Effect on unemployment

Fig 7.2 shows the effect on unemployment when this method of control was used on the LBS model (Zarrop et al, [1]). It is of interest to note the difference in the effectiveness of the control in the open-loop and closed-loop implementations, especially as the linear model was not derived from the large model but estimated directly from economic data. It would appear that the feedback compensates reasonably well for the model mismatch in this case.

7.6 ITERATIVE RESPECIFICATION OF THE QUADRATIC COST FUNCTION

A fundamental problem in the optimization of policy decisions is the specification of a suitable objective function. Even if we consider only the class of quadratic objective functions, we are still faced with the problem of choosing the weighting matrices. Of course, this is not the policymaker's problem. He is 'only' interested in a politically acceptable path in line with the government's objectives. Rustem et al [13] have devised a method that generates such a path using the optimization framework.

Let F be the set of trajectories $Z = (Y^T, U^T)^T$ that satisfy the model constraints and Ω the set of trajectories that are acceptable to the policymaker. The problem is to generate a point in the intersection of F and Ω (assuming that it is not empty) by minimizing the cost

$$J(Z) = \frac{1}{2} < \delta Z, Q \delta Z > , \qquad (7.11)$$

with a suitable positive definite symmetric matrix Q.

Let Q_c denote the current weighting matrix and Z_c (in F) denote the optimal trajectory corresponding to Q_c. If this trajectory lies in Ω the policymaker is satisfied. If not, the policymaker is asked to criticise the solution and put forward a preferred trajectory Z_p. Introduce the correction vector δ given by

$$\delta = Z_p - Z_c , \qquad (7.12)$$

and a vector γ given by

$$\gamma = \nabla J_{true}(Z_p) - \nabla J_{true}(Z_c) \qquad (7.13)$$

measuring the difference in the gradients of the true (unknown) cost function at the current optimal and preferred trajectories. A new weighting matrix Q_n giving a new optimal solution nearer to the preferred values may be computed by adding a rank one correction term to Q_c

$$Q_n = Q_c + \frac{(\gamma - Q_c \delta)(\gamma - Q_c \delta)^T}{\langle (\gamma - Q_c \delta), \delta \rangle} \quad (7.14)$$

(This updating formula is a member of the variable metric family discussed in chapter 1.3, and widely used for computing approximations to second derivative matrices in both constrained an unconstrained optimization.) The matrix Q_n is now used to generate a new optimal solution Z_n, and so on. Fig 7.3 illustrates the iterative nature of the method.

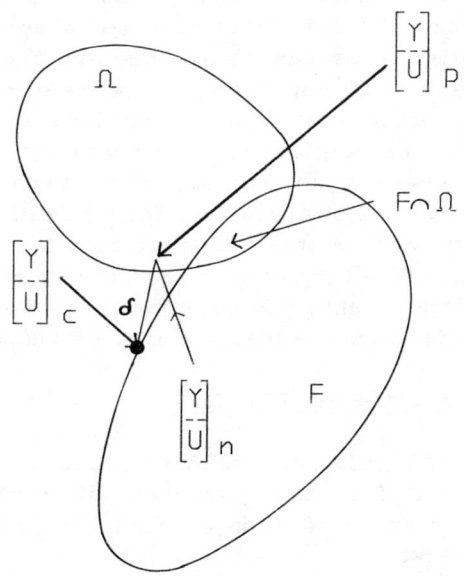

Fig 7.3   Iterative respecification of weighting matrix

The motivation for convergence of the algorithm is a theorem of Fiacco and McCormick [9] proving that if the true objective function is a positive definite quadratic form then the updates converge to the true weighting matrix in a finite number of iterations no greater than the dimension of Z, provided the generated correction vectors are linearly independent. Of course, it is assumed that the specified δ and γ vectors are consistent with the true objective function, and we cannot rely on the policymaker to satisfy this requirement.

Choosing γ is the main obstacle to implementing the algorithm. A useful device that works well in practice is to choose

134 Control theory in macroeconomics

$$\gamma = \phi Q_c \delta \, , \qquad (7.15)$$

where ϕ is a scalar greater than unity. Then (7.14) becomes

$$Q_n = Q_c + (\phi-1)\frac{(Q_c\delta)(Q_c\delta)^T}{(\delta^T Q_c \delta)} \, . \qquad (7.16)$$

The larger the magnitude of ϕ in a particular iteration, the more emphasis the policymaker places on achieving his preferred trajectory in that iteration.

It should be noted that, in general, the updates introduce intertemporal and intervariable cross-weightings, e.g. connecting the tax rate in 1970(1) with unemployment in 1974(3). Intertemporal weightings can be avoided and the cost function form (7.7) preserved if corrections are made to trajectories at only one time period at each respecification.

The interactive nature of this respecification technique enables the policymaker to explore the consequences of his preferences and to learn more about the economic model and the kinds of hard economic constraints that face him when decisions have to be made. Optimality becomes a means to this end, rather than an end in itself and, in this context, there is clearly no need to consider non-quadratic forms of the cost function.

7.7 CONTROLLING A LARGE MODEL: II

The task of directly controlling large and nonlinear econometric models for the purposes of economic policy evaluation has led to the design of a number of algorithms, of which we mention two.

Consider the cost function (7.7) in the static form

$$J(Y,U) = \frac{1}{2}\langle \delta Y, Q_y \delta Y \rangle + \frac{1}{2}\langle \delta U, Q_u \delta U \rangle \, , \qquad (7.17)$$

where intervariable terms are omitted for simplicity. We write the $U \to Y$ mapping (generated by the model solution program) as

$$Y = g(U) \, . \qquad (7.18)$$

Writing

$$G(U) = J(g(U),U) \, , \qquad (7.19)$$

the optimization is now unconstrained, i.e.

$$\min_{U} G(U) \:. \tag{7.20}$$

Note that this problem can be of high dimension. For example, if we select five instruments over five years (20 periods), we have 100 optimisation variables.

One method of solving (7.20) is to employ a Newton-type descent algorithm with variable step length (Rustem and Zarrop [12]), of the form

$$U \rightarrow U - \alpha H^{-1} \nabla G(U) \:, \tag{7.21a}$$

where

$$\nabla G(U) = N^T Q_y (g(U) - Y^d) - Q_u (U - U^d) \:, \tag{7.21b}$$

and the positive definite matrix

$$H = N^T Q_y N + Q_u \:, \tag{7.21c}$$

are the gradient and (approximate) Hessian respectively of $G(U)$, the model Jacobian N is given by

$$N_{ij} = \frac{dY_i}{dU_j} \:, \tag{7.21d}$$

and the step length α is chosen to ensure a sufficient decrease in the cost at each iteration. If the model is linear, the algorithm converges in one step with $\alpha = 1$.

In general, it is too onerous a task to generate the Jacobian analytically, yet for a nonlinear model the alternative appears to be to generate the derivatives numerically at each iteration. This would involve, say, 100 extra model runs at each iteration. Two approximations can be made that reduce this computational load to an initial number of runs equal to the number of instruments.

Initially, it is assumed that the elements of N have the Markov matrix form

$$\frac{\partial y_i(m)}{\partial u_j(k)} = M_{ij}(m-k), \quad k \leqslant 1, \tag{7.22}$$
$$= 0 \quad \text{otherwise,}$$

which holds for a causal, time-invariant, linear model. Then N can be built up numerically by perturbing each instrument independently in the first time period only and takes the banded form

$$N = \begin{bmatrix} M(0) & 0 & \cdots\cdots & 0 \\ M(1) & M(0) & \cdots\cdots & 0 \\ \cdot & \cdot & & \cdot \\ \cdot & \cdot & & \\ M(N-1) & M(N-2) & \cdots\cdots & M(0) \end{bmatrix} \cdot$$

(7.23)

At each iteration N can be corrected by a rank one update (Broyden, [6]) without additional runs, although this destroys the banded lower triangular structure.

An alternative approach (Chow, [7]) is to linearise the model and solve an LQ problem in feedback form at each iteration. This involves the undesirable initial task of casting the nonlinear model in state space form. More seriously, convergence is not guaranteed. The algorithm is of the Newton type with unit step length and thus may jam up, particularly if the initial trajectory is not close to the optimum.

7.8 CLOSED LOOP v OPEN LOOP

The generation of optimal state feedback laws such as (7.9) that arise naturally in the Chow algorithm appears to be of limited use in economic policy formulation. Models, data, exogeneous forecasts and even priorities can change rapidly and demand frequent policy revisions. The Treasury, for example, holds two major short-term forecasting rounds each year, during which policy options are generated based on the current model and information. This round of activity is perhaps most closely reflected in the sequential open-loop approach of Athans et al [4] in which the policy maker generates an optimal policy at time i for periods $i+1$, $i+2$,...,$i+N$ but only implements this policy for the first few periods before repeating the process on updated information.

On the other hand, feedback laws in the form of simple, robust rules of thumb connecting, say, changes in government spending and the money supply with changes in the levels of prices and unemployment are clearly of interest as a guide to possible stabilisation policies.

7.9 POLICY SENSITIVITY

Returning to the general stochastic model

$$Y = f(Y,U,E,\theta) + \varepsilon ,$$

(7.23)

what are the possible effects on policy of the errors arising from E, θ, ε? Where the errors enter as additive linear terms (e.g. linear model with no parameter error) then certainty equivalence solves the problem of minimizing the mean of the quadratic cost by telling us to solve the deterministic problem with the random variables replaced by their means. Where nonlinearity occurs, we enter the realms of nonlinear stochastic control theory, but even if a solution is derived it may be risky to use it in the sense that the variance of the cost at the optimal trajectory may be large. Clearly, in the presence of uncertainty, we should pursue more cautious (risk averse) policies.

We take a 'sensitivity' rather than a 'stochastic' approach. We wish to minimize

$$J(Y,U) = \frac{1}{2} < \delta Y, Q_y \delta Y > + \frac{1}{2} < \delta U, Q_u \delta U > , \qquad (7.24)$$

subject to the linear (or linearised) model

$$Y = NU + c + \tilde{N}\omega , \qquad (7.25)$$

where ω is error from all relevant sources. Denote by ξ the sensitivity vector

$$\xi = (\partial J/\partial \omega)|_{\omega=0} = \tilde{N}^T Q_y (\delta Y)|_{\omega=0} \qquad (7.26)$$

We can penalise trajectories of high sensitivity by augmenting the cost by a term proportional to $<\xi, \Sigma\xi>$ where Σ is a suitable symmetric positive definite weighting matrix. We now pose the problem as one of minimizing

$$\hat{J}(Y,U) = J(Y,U) + \frac{1}{2}\beta < \xi, \Sigma\xi >$$

$$= \frac{1}{2} < \delta Y, \hat{Q}_y \delta Y > + \frac{1}{2} < \delta U, Q_u, \delta U > , \qquad (7.27)$$

where

$$\hat{Q}_y = Q_y + \beta Q_y \tilde{N} \Sigma \tilde{N}^T Q_y \qquad (7.28)$$

subject to the error-free model

$$Y = NU + c . \qquad (7.29)$$

The form of the problem is that of the standard LQ type, but in general $\tilde{N}$ is not constant. It is therefore suggested that a series of optimizations is carried out starting with zero β and increasing it by a <u>small</u> amount for each subsequent optimization. Then $\tilde{N}$ can be held constant for each optimization without excessive error. The process is terminated when a satisfactory trade-off is obtained between the original objectives and reduction in sensitivity. The results can be checked by Monte Carlo simulation.

References

A number of the references appear in:

[1] Optimal Control for Econometric Models: An approach to economic policy formulation, S. Holly, B. Rustem & M. B. Zarrop (Eds), Macmillan.

Also of interest are:

[2] Report of Committee on Policy Optimization, Cmnd. 7148, March 1978, HMSO;

and the review article:

[3] Kendrick, D. (1976): Applications of Control theory to Macroeconomics, Annals of Economic and Social Measurement, 5, 2, pp. 171-190.

Other references:

[4] Athans, M. et al (1976): Sequential open-loop optimal control of a nonlinear macroeconomic model in Frontiers of Quantitative Economics, M.D. Intriligator (Ed), North Holland.

[5] Box, G.E.P. & Jenkins, G.M.(1970): Time Series Analysis, Forecasting and Control, Holden Day.

[6] Broyden, C.G. (1965): A class of methods for solving nonlinear simultaneous equations, Mathematics of Computation, 25, 223-245.

[7] Chow, G.C. (1975): Analysis and Control of Dynamic Economic Systems, Wiley.

[8] Derakhshan, M. (1978): Bibliography on Application of Systems and Control Theory in Economic Analysis, Oxford Eng. Lab. Rpt. 1267/78.

[9] Fiacco, A.V. & McCormick, G.P. (1968): Nonlinear Programming: Sequential Unconstrained Minimization Techniques, Wiley.

[10] Holly, S., Zarrop, M.B. & Westcott, J.H.(1978): Near-optimal stabilisation policies with large nonlinear econometric

models : an application to the LBS econometric model of the U.K. economy, Imperial College, CCD Publication 78/12.

[11]   Preston, A.J. & Wall, K.D. (1973): An extended identification problem for state space representations of econometric models, PREM Discussion Paper 6, CCD, Imperial College.

[12]   Rustem, B. & Zarrop, M.B. (1979): A Newton-type method for the optimization and control of nonlinear econometric models, Journal of Economic Dynamics and Control, 1, 3, pp. 283-300.

[13]   Rustem, B. Velupillai, K. & Westcott, J.H. (1978): Respecifying the weighting matrix of a quadratic objective function, Automatica, 14, pp. 567-582.

[14] Wall, K.D. (1976): FIML estimation of rational distributed lag structural form models, Annals of Economic & Social Measurement.

[15] Wall, K. D. & Westcott, J. H. (1974): Macroeconomic modelling for control, IEEE AC-19, 6, pp.862-873.

Chapter 8
Optimal operation of canal reservoirs
P. Nash

8.1 INTRODUCTION

The British Waterways Board (B.W.B.) manages a network of about 2,400 km of canals, about 400 km of river navigation and 90 reservoirs. The canals were built mainly in the late 18th. and early 19th. centuries, for the passage of freight. The invention of the railway led to a sharp decline in freight traffic, which has continued to the point where very little freight traffic now exists, and the canals are used almost exclusively for recreation. Pleasure boating in particular has seen a very rapid growth in recent years.

The prime responsibility of B.W.B. in managing the system is to keep the canals and reservoirs in good repair and ensure that they contain enough water for navigation. However, sales of water contribute substantially to income, and so the ability to supply water to industrial users is an important consideration. In addition, many reservoirs have important amenity value, either as scenic features or as bases for sailing and angling clubs, and it is desirable that the water levels in these reservoirs are not allowed to become too low.

Recent events have combined to persuade B.W.B. that an analysis of the problem of reservoir control is necessary. First, a succession of very dry years has shown up limitations in the present methods of control. Secondly, the increased pleasure boat traffic has caused an increased demand for water. It may seem at first sight surprising that boats use up water. This happens because canals are routed as much as possible along contours of the land, but must occasionally climb or descend. This is achieved by using locks to connect stretches of canal at different levels, and when a boat passes through a lock, a lockful of water - up to 100,0001. - passes from the higher to the lower level. This has to be replenished at the higher level, and the canal is kept full by feeding in water from the reservoirs, usually into the highest or summit level. Thirdly, deterioration of the reservoirs with time has meant that some of them cannot be operated at full capacity. This makes the good management of what capacity does exist essential.

Thus B.W.B. are faced with the need to decide how much increase in boat traffic can be supported on individual canal systems, and how far deteriorating reservoirs have to be improved to ensure adequate supplies of water. The answers to these questions depend on how well the reservoirs can be controlled: better control will enable a given system to support a greater demand, or enable a given demand to be met from a smaller reservoir capacity. This chapter describes a model of the reservoir control problem which has been developed with the aim of providing improved control rules for week to week use on the actual system, as well as providing a means of investigating these longer term questions of strategy.

The approach described here makes use of the special features of a canal system to provide very simple and easy to implement control rules. In the next section, we describe an optimizing model that works by breaking the year into two periods and combining near-optimal control rules for each. We then give examples to show how these rules have been implemented on the Caldon canal.

8.2 MODELLING RESERVOIR SYSTEMS

If all we are interested in is the water level, then a reservoir is about as simple a dynamic system as we can imagine. It is described by the state equation

$$\dot{x}(t) = v(t) - u(t)$$

Complications arise because, in general, the inflow $v(t)$ is stochastic, the outflow $u(t)$, which is the control, has to be chosen to meet a demand which is unknown ahead of time, and because the contents $x(t)$ are constrained to be non-negative and less than the reservoir capacity. The problem we face in modelling the reservoir control problem is to decide the level of detail at which to model this complication so as to get useful results.

A large amount of work on problems of reservoir control has been reported by numerous authors, including Fults and Hancock [1], Harboe et al[2], Kindler [3], Maass [4], Moran [5], Reinisch et al [6], Roefs and Boden [7], Russel [8], Schweig and Cole [9], and Stephenson [10]. Almost none of this work would be directly applicable to the B.W.B. system for various reasons, chief among which are that:

(i) it presupposes a much more complex system of data collection and control than is available to B.W.B.,

(ii) it does not include one very important feature of B.W.B. reservoirs, namely that the rate at which water can be fed out from a reservoir may be subject to quite severe constraints, and

(iii) it does not allow sufficient flexibility to take account of changed circumstances without re-doing complicated calculations.

For example, a common approach is something like the following. Let

$$J(x_0,u) = \sum_{t=0}^{T} I_t(x_t, v_t, u_t) ,$$

where x_t is the vector of reservoir holdings at time t, v_t is the vector of inflows and u_t the vector of outflows at that time, and I_t is some function of these variables. We choose the form of the functions $\{I_t\}$ to reflect the notional cost of such events as failure to meet demand, low water or flooding in the reservoirs, and use an optimization technique to minimize the objective functional J subject to the constraints imposed by the system characteristics. The system dynamics tend to be linear, so choosing the functions $\{I_t\}$ quadratic will lead to a linear-quadratic control problem. This can be attacked in a deterministic form, assuming a specific (either historical or synthetic) set of values for all the inflows and demands, as in for example Harboe et al [2], or in a stochastic form, using a probabilistic model for inflows and demands, as in Reinisch et al [6]. The former approach is perhaps most useful in the assessment of stategical alternatives, while the latter will result in on-line control strategies.

The first of these could be useful in the context of canal reservoir systems, but is handicapped by the computational requirements of the approach. The second suffers from this disadvantage, and is further unsuitable because of the sophistication of the control schemes produced. Neither approach is likely to lead to simple rule-of-thumb methods for improving reservoir control, and the level of mathematical sophistication involved makes it unlikely that water engineers would gain any appreciable new insight into the operation of their reservoir systems. Of course, none of this is to say that these approaches cannot be of use in other cicumstances.

8.2.1 A simple optimal control model

A special feature of canal reservoirs is that the demand they experience is highly seasonal, with a base level of demand varying only slowly over the year, and a well defined period of much increased demand due to boat traffic during the period from late spring to early autumn. The base demand is water that has to be fed into the canal to make up for industrial abstraction, seepage and evaporation (the last two of which also have their maxima in the summer). Hence the year can be divided into two periods: a period during which reservoir levels are falling, and another during which they are rising. Waterways engineers recognise this, and refer to the two periods as "drawdown" and "refill" periods. As a first step, therefore, we might consider three separate questions:

(i)   what is the optimal control during the drawdown period?

(ii) what is the optimal control during the refill period?

(iii) when should control switch from one to the other?

Consider a system of n reservoirs, $R_1, R_2, \ldots R_n$. The holding of R_i at time t is denoted by $x_i(t)$, and the amount of space remaining in R_i by $y_i(t)$. The total capacity of R_i is C_i. Thus for each reservoir R_i,

$$x_i(t) + y_i(t) = C_i.$$

The rate of runoff into R_i at time t is in general a random variable, and is denoted by $r_i(t)$. At each time t there is a demand to be supplied from the reservoirs; again this is in general a random variable, and it is denoted by $d(t)$. The rate of feed from reservoir i at time t is denoted by $u_i(t)$. These are constrained by feeder capacities:

$$u_i(t) \leqslant U_i.$$

We shall denote by $x(t)$ the vector $(x_1(t), x_2(t), \ldots, x_n(t))^T$ and by $u(t)$ the vector $(u_1(t), u_2(t), \ldots, u_n(t))^T$.

8.2.2 Optimal control on drawdown

Consider the drawdown of the reservoirs during the summer months. During this period runoffs of water into the reservoirs are small. We disregard them and consider the problem of satisfying as much as possible of a demand $d(t)$ up to some time horizon T, in the absence of runoff. This objective does not completely encapsulate the aims of the engineers in controlling the reservoirs during drawdown. In particular, amenity considerations are ignored, and this will have to be borne in mind when interpreting and implementing any solution of this simplified problem. Nonetheless, it seems likely that a policy which achieved this simplified objective would give good results in practice.

Suppose for the moment that the demand $d(t)$, $0 \leq t \leq T$, is a known function of time. Then our problem can be stated formally as

$$P: \text{ maximize } \int_0^T \left(\sum_{i=1}^n u_i(t) \right) dt$$

$$\text{subject to } \dot{x}_i(t) = -u_i(t), \quad i = 1, 2, \ldots, n$$

$$x_i(t) \geq 0, \quad i = 1, 2, \ldots n$$

$$0 \leq u_i(t) \leq U_i, \quad i = 1, 2, \ldots, n$$

$$\sum_{i=1}^n u_i(t) \leq d(t)$$

P would be a trivial optimal control problem, were it not for the state constraints $x_i(t) \geq 0$. The presence of these means that the minimum principle as discussed in chapter 2.3.1 does not immediately apply. A discretization of P would be a linear program, and hence soluble by the simplex algorithm. This would not be of much value here, since we are seeking insight and general principles rather than a specific solution to the rather unrealistic problem P. Instead, we approach the solution of P by a combination of dynamic programming and ad hoc methods.

Let $V(x,t)$ be the maximum amount of the demand that can be met over the interval $[t,T]$, if the reservoirs are in state x at time t. Then the optimality principle says

$$V(x,t) = \max_{\Omega} \left\{ \delta \sum_{i=1}^n u_i + V(x - u\delta, t + \delta) \right\} \qquad (8.1)$$

146 Optimal operation of canal reservoirs

for δ small enough, where

$$\Omega = \{u: u \geqslant 0, \ \Sigma u_i \leqslant d(t), \ u_i = 0 \text{ if } x_i(t) = 0\}.$$

Letting $\delta \to 0$, we have

$$\frac{\partial V}{\partial t} = -1 + \max\{-\sum_{i=1}^{n} u_i(1 + \frac{\partial V}{\partial x_i})\}. \tag{8.2}$$

Now clearly, $\partial V/\partial x_i \geqslant 0$ for all i, since adding water to any reservoir cannot decrease V. Moreover, if for an optimal trajectory $x_i(T) > 0$, then adding water to R_i cannot increase V either. For such a reservoir, therefore, $\partial V/\partial x_i = 0$ throughout [0,T]. The state constraint never operates, so $-\partial V/\partial x_i$ behaves as the co-state variable would normally: the boundary condition at time T says that it is zero then, and since there is no explicit dependence on x_i in either objective or dynamics, it is constant throughout [0,T].

From these considerations, we can deduce that an optimal strategy will never take water from the first reservoir to empty unless it is unavoidable. This will occur only at times when the demand is at least as great as the combined feeder capacities of the other reservoirs. Consider a demand function d which is always at least as big as any (n-1) combined feeder capacities, and suppose for definiteness that under an optimal drawdown strategy, R_1 empties strictly earlier than any other reservoir. Then the other reservoirs will feed at maximum rates throughout [0,T]. It follows that

$$\frac{x_i(0)}{U_i} > \frac{x_1(0)}{U_1}$$

for all $i > 1$. Thus the reservoir that empties first under these circumstances is one with the smallest value of x_i/U_i. It is straightforward to show that this remains true for smaller demand functions.

We define the drawdown time D_i of R_i at time t by

$$D_i(t) = \frac{x_i(t)}{U_i}.$$

It is the time required to empty R_i if water is fed out at the maximum feeder rate. The greatest drawdown time policy (GDT) is defined as the policy which as far as possible always supplies the demand from the reservoir whose drawdown time is currently greatest. Thus under GDT, if

$$\frac{x_i(t)}{U_i} > \frac{x_j(t)}{U_j},$$

then $u_j > 0$ only if $u_i(t) = U_i$. If the greatest drawdown time is shared by two or more reservoirs, the feeds are chosen to maintain this equality.

The action of the GDT policy is the intuitively appealing one of trying to arrange the reservoir contents so that

$$\frac{x_i}{(x_1+x_2+\ldots x_n)} = \frac{U_i}{(U_1+U_2+\ldots U_n)}$$

for each reservoir R_i. It is also completely independent of future values of d, and optimal for any demand pattern. Two generalizations follow from these latter remarks. First, GDT maximizes the time for which any demand can be met. For suppose that a demand $d'(t)$ can be met throughout $(0,T')$ by a control u. Then GDT must supply at least as much water as u over the interval $(0,T')$, and hence also meets the demand.

The second point is that, since GDT is optimal for any demand pattern, and independent of future values of demand, it must be optimal for an unknown or stochastic demand. This follows from the fact that for such a demand, even if we knew future values, we would make no use of that knowledge. The greatest of the simplifications made in formulating P - that demand is known - therefore does not affect the optimal drawdown policy.

8.2.3 Optimal control on refill

We now consider the problem of controlling the reservoirs during refill so as to catch as much of the runoff as possible. Implicit in this objective is an assumption that water is of equal value wherever it is caught. The drawdown model indicates that this may well not be so, in that the feeder constraints may imply that storage above a certain level in some reservoirs may be of little or no value. Again, though, it is plausible to suggest that a policy based on optimizing this objective should be a good one in practice. The demand during this period is relatively small so we disregard the feeder constraints, which are not usually active. We define the "refill time" of R_i as the time at which the reservoir would fill if no water were fed out. Clearly, this depends on future runoff. The "least refill time policy" (LRT) is the policy of drawing on the reservoirs which always supplies the demand from the reservoir whose refill time is the least. LRT can be implemented for known future runoff, and arguments like those used to prove optimality of GDT can be used to show that it is optimal in that case.

Implementation of the GDT policy for drawdown would be relatively straightforward, since the control actions determined by the policy can be calculated from a knowledge of the reservoir contents and the feeder constraints. The same is not true of LRT, because of its dependence on the unknown future runoff, at least as it stands. However, the reservoirs of a canal are usually located near to one another at the canal summit. Because of this they experience about the same pattern of rainfall, and the respective runoffs are in approximate proportion to their catchment areas.

If we take this to be so, then the refill time of a reservoir, as we have defined it, is proportional to the empty space remaining in the reservoir. Thus if the catchments of the reservoirs are in the ratios $k_1:k_2:\ldots k_n$ then their refill times are in the ratios $y_1/k_1:y_2/k_2:\ldots y_n/k_n$, and it is now possible to compute the control actions determined by LRT from known information. The action of LRT is to try to arrange things so that

$$\frac{y_1}{k_1} = \frac{y_2}{k_2} = \ldots = \frac{y_n}{k_n}.$$

8.2.4 A Two Regime Policy

We have seen that there are two simple rules, GDT and LRT, which might be expected to perform well for drawdown and refill respectively. The policy we propose for reservoir control throughout the year is one which during any week acts as either GDT or LRT. The decision as to which to use may involve the time of year, the state of the reservoir holdings, the runoff pattern and the level of the demand.

The simplest policy, which we shall denote by D/R (drawdown/refill), is one based completely on the time of year. Weeks of the year are labelled 1-52. Two weeks are selected,

dw_1 and dw_2. The GDT rule is applied during the "drawdown weeks", dw_1 to dw_2 inclusive, and LRT is applied during other weeks. The values of dw_1 and dw_2 are the same for all years and are chosen by computer simulation of previous difficult years.

In practice, one might be able to choose between operating a drawdown or refill policy more suitably by taking other factors into account. To help the engineer to do this we introduce another policy D/R*. This rule has parameters, rw_1, rw_2, dw_1, dw_2, d^* and w^*. The LRT rule is always applied during the "refill weeks", rw_1 to rw_2 inclusive. The GDT rule is always applied during the weeks dw_1 to dw_2 inclusive, but with the proviso that LRT is applied instead if the present holdings are sufficient to certainly meet a demand d^* as far as the "target week", w^*. During weeks not covered by rw_1-rw_2 or dw_1-dw_2 the LRT or GDT rule is applied as the expected runoff that week is greater or less than the demand (the expected runoff might be taken as runoff of the previous week). The same proviso about meeting the maximum demand as far as the target week applies.

The parameters of the D/R* policy all have simple interpretations, and can be chosen suitably by examining historical records. The period of the refill weeks can be taken as the period during which every one of the historical years under study could be said to have been on refill (it will be shorter period than the refill period of a particular historical year). Similarly the drawdown weeks can be taken as those weeks which define a period when every historical year could be said to have been on drawdown. The maximum demand, d^*, can be chosen to reflect the demand rate expected if the drawdown period is prolonged, and the target week, w^*, can be set at one week before rw_1.

The rules GDT and LRT have the advantage that the calculations of x_i/U_i and y_i/k_i required to implement them are simple. It is also relatively easy to determine whether the holdings are sufficient to satisfy a demand of d^* as far as week w^*.

8.3 CONTROL OF THE CALDON RESERVOIRS

The Caldon canal is 280 km long, running from Etruria to Froghall, near Stoke-on-Trent. It has three reservoirs: Knypersley Pool (R_1), Rudyard Lake (R_2) and Stanley Pool (R_3), which have capacities of 1155, 3455 and 611 Ml respectively. A further source of water is treated effluent, pumped from the town of Stoke. The feeders from the three reservoirs are subject to capacity constraints which are currently 50, 105, and 160 Ml/wk respectively.

8.3.1 Assessing the proposed control rules

Implementation of the D/R and D/R* policies has been computer simulated for the three reservoirs of the Caldon canal during 1974, 75, and 76. These years, being the driest run of years this century are considered to be a good basis for evaluating the affect of control rules, or changes in feeder constraints or demands. The week by week runoffs and demands actually recorded during these years were used in the simulation. The assumption of proportional runoffs was checked by simply noticing that the runoffs to the three reservoirs were roughly the same week by week, and by simulating a policy which implemented an exact least refill time policy. It was found that the result of this simulation differed little from one which was based on proportional runoffs. Values of k_1, k_2, and k_3 were estimated from the three years' records as 6, 7, and 6 respectively.

Table 1 gives the reservoir contents for various weeks of 1974-76 when feeder rates are controlled by the D/R policy with $dw_1 = 15$ and $dw_2 = 34$. Table 2 gives the results of control with the D/R* policy, using parameters determined as described above: $rw_1 = 47$, $rw_2 = 11$, $dw_1 = 20$, $dw_2 = 34$, $d^* = 160$, $w^* = 46$. The D/R* policy is superior: it takes less risk during the summer and ends the year 1976 with marginally greater total holdings (4750 as against 4708 Ml). In actual practice B.W.B. ended 1976 with total holdings of 4398 Ml. They reached a low of 1284 Ml during the drought of 1975 as compared to 1694 Ml for the D/R* policy.

TABLE 1   D/R policy

Year	Week	Kny	Rud	Sta	Total
1974	1	1155	3455	611	5221
	35	756	1336	454	2546
1975	1	1155	3455	611	5221
	46	343	1369	0	1712
1976	1	1155	2610	561	4327
	38	509	1183	75	1768
	52	1155	2942	611	4708

The first column gives weeks of 1974, 75, and 76 chronologically. The second to fourth columns give reservoir holdings in those weeks (Ml). The fifth column gives total holdings.

TABLE 2   D/R* policy

Week	Contents			Total
	Kny	Rud	Sta	
1	1155	3455	611	5221
35	642	1859	44	2546
1	1155	3455	611	5221
46	454	1239	0	1694
1	1155	2480	587	4223
38	316	1225	96	1639
52	1155	2984	611	4750

These simulations comfirm the efficacy of the D/R* policy, and as a result its implementation in practice was recommended. This recommendation was accepted, and some simple control charts were designed to enable the rule to be put into practice.

8.3.2 Two strategical questions

The possibility for improved operation provided by this rule suggested that it might be feasible to run the system without Stoke effluent during certain parts of the year. This is an attactive possibility because it avoids pumping costs and the degradation of water quality caused by the current intake of 80 Ml/wk of effluent.

Table 3 shows that by using the D/R* rule, Stoke effluent pumping can be eliminated for weeks 47-13 inclusive. Even in the severe drought years 1974-76 circumstances would not have been significantly worsened if pumping had been so reduced.

TABLE 3 Reduced effluent pumping

Week	Contents			Total
	Kny	Rud	Sta	
1	1155	3455	611	5221
35	642	1735	43	2421
1	1148	3436	595	5180
46	458	1220	0	1679
1	1113	2340	326	3780
38	316	1085	96	1499
52	1002	2844	380	4227

Another question that was considered was that of what improvements in the feeder capacities are necessary to deal with a greatly increased demand pattern. A demand pattern is

TABLE 4 A projected demand with current feeder constraints

Week	Contents			Total	Fail-ures
	Kny	Rud	Sta		
1	1155	3455	611	5221	
35	770	1534	611	2916	
1	1155	3455	611	5221	
25	969	2857	342	4168	33
26	926	2709	194	3830	23
27	867	2555	48	3471	32
28	833	2406	0	3239	15
46	660	1497	141	2299	
1	1155	2738	611	4504	
28	786	2185	88	3060	2
29	716	2045	0	2761	62
30	646	1909	0	2556	190
31	615	1778	0	2394	200
32	547	1646	0	2194	158
33	463	1516	0	1980	196
34	379	1361	0	1740	154
52	1155	2888	611	4654	

projected for increased industrial extraction and boat traffic in the 1980's. Table 4 shows that with present feeder constraints this demand could not be met without failure. By increasing the feeder rates of Rudyard and Knypersley to 160 and 90 Ml/wk the demand can be met, as shown in Table 5. In fact the Stanley feeder need only be capable of 130 Ml/wk. If a less good policy is employed then the feeder capacities must be larger. The sixth column gives the amount by which the demand fails to be met (Ml).

TABLE 5   A projected demand with improved feeders

Week	Contents			Total
	Kny	Rud	Sta	
1	1155	3455	611	5221
35	829	1475	611	2916
1	1155	3455	611	5221
25	931	2675	528	4135
26	848	2472	452	3773
27	749	2263	369	3382
28	681	2059	394	3135
46	634	1144	416	2195
1	1155	2385	611	4151
28	751	1452	501	2705
29	642	1256	445	2344
30	532	1066	350	1949
31	461	879	244	1586
32	353	693	181	1227
33	229	508	80	817
34	105	297	20	423
38	36	65	3	105
52	1034	1824	611	3469

8.4 APPLICATION TO OTHER SYSTEMS

The approach described here has since been followed in investigations of other systems, notably the Macclesfield branch of the Caldon-Macclesfield system and the Leeds and Liverpool canal. Precisely because the methods used depend so strongly on the special characteristics of the system, the modelling part of the exercise has to be gone through afresh in each case. Modifications and extensions to the basic results have had to be made to cope with different circumstances. In the case of the Leeds and Liverpool canal, demand cannot be treated as a single number, since parts of the demand can be met from only a subset of the reservoirs. In other cases, the possibility of moving limited amounts of water from one reservoir to another has to be taken into account. Further complications can arise if back-pumping of water from lower to higher levels is possible, as it is at a number of places, or if water from, say, boreholes can be used as well as supplies from reservoirs. These last two introduce monetary cost as another important criterion of efficient operation, because of the high cost of pumping.

In these more complicated situations, solution of models even as simple as those presented here is difficult, and it becomes more and more necessary to use intuition – still founded on consideration of the simple models – to bridge the gap between the theory and practical heuristics for control. In doing this, the use of simulation becomes crucial, as a demonstration to the engineer of the likely performance of any specific strategy, and as a means of enabling him to assess the trade-off between the different criteria he has to consider. Current work is centred on the implementation in a micro-computer of a general simulation program, which will allow the engineer to build a model interactively, incorporating any of the water storage and transport elements occurring in the real system, and using historical runoff data.

It has also become clear that it is of great importance to obtain the best possible estimates of the likely increases in demands consequent on increased numbers of boats. Work in this area is under way, and one ultimate goal is to be able to relate directly the effects of improved control or upgrading of installations to the extra boat traffic that can be supported thereby.

REFERENCES

A good source of material on the latest research in applying systems modelling and optimization techniques to planning and control of water resource systems is the Proceedings of the IFAC Symposium on Water and Related Land Resource Systems, Cleveland, 1980, from which two of the references below come.

[1] Fults, D.M. and Hancock, L.F. (1972). Optimal operations model for the Shasta-Trinity system, J. Hydr. Division, Am. Soc. Civ. Eng. 98, 1497-1514.

[2] Harboe,R., Schultz,G.A., and Duckstein,L. Low-flow and flood control:distributed versus lumped reservoir model Proc. IFAC Symposium on Water and Related Land Resource Systems, Cleveland, 1980.

[3] Kindler,J. (1975) The Monte-Carlo approach to optimization of the operating rules for a system of storage reservoirs. Proc. Int. Symp. on Appl. of Math. Models in Hydrology and Water resource Systems, Bratislava, 1975

[4] Maass,A. (1962). Design of Water Resources Systems, Macmillan, New York.

[5] Moran, P.A.P. (1959). The Theory of Storage, Methuen and Co. Ltd., London.

[6] Reinisch,K., Irmscher,S., and Thumler,C. (1980) Application of one-level and multilevel methods for developing control strategies for a special water resources system Proc. IFAC Symposium on Water and Related Land Resource Systems, Cleveland, 1980.

[7] Roefs, T.G. and Bodin, L.D. (1970). Multi- reservoir operation studies, Water Resources Research, 410-420.

[8] Russel, C.B. (1972). An optimal policy for operating a multi-purpose reservoir, Opnes. Res. 20, 1181-1889.

[9] Schweig, Z. and Cole, J.A. (1868). Optimal control of linked reservoirs, Water Resources Research, 4, 479-497.

[10] Stephenson, D. (1970). Optimal design of complex water resource projects, Proc. Am. Soc. Civ. Engrs., 1229-1246.

Chapter 9
Coal market modelling
R. A. Blewitt

9.1 INTRODUCTION

The Operational Research Executive (ORE) of the National Coal Board makes surprisingly little use, for an organisation of its type and size, of linear programming (LP). Although LP is used on a limited scale by ORE, it has not achieved the position of importance that it occupies in the literature of OR or in some other OR groups. There are various reasons for this, and this chapter will look at some of them, in particular those related to using LP for marketing problems. The paper describes the construction of a market model for Barnsley Area of the National Coal Board, and why LP is being used again for this type of problem, despite having been found unsuitable in previous attempts.

The first section of this chapter describes the background to the present work, examining the problems of coal marketing and earlier attempts in this field. Then the current work is looked at, and finally the results of this recent work described.

9.2 BACKGROUND

Coal is not of uniform quality; it varies in size, calorific value, dirt content, coking properties, and sulphur content. There are a variety of markets it can be sold to: domestic, C.E.G.B, coking, industrial. The price of coal in each market is different and depends on its quality.

To increase the price obtained for coal in these markets, coal is 'washed' (to remove dirt) and segregated by size in coal washeries or preparation plants. Further it is blended to produce mixes of suitable quality for the appropriate markets. This is referred to as coal 'preparation'.

9.2.1 Barnsley Area

The National Coal Board is divided into 12 Areas. Each Area, covering a geographical group of pits, forms a semi-autonomous unit responsible for mining and marketing the coal located within its Area boundaries (subject to overall national guidance and constraints). Barnsley Area consists of 18 pits (centred roughly on Barnsley town) with a total annual output of about 7 million tons in 1978/79. The area has total classified reserves of coal of 300 million tons, and 28 different seams. Barnsley Area has 61% of total National reserves of very strongly coking coal (technically coal of rank 400), with coking coal making up 82% of the Area's proved reserves. (For further information see 'Barnsley Area in the 1980's' - Mining Technology February 1980).

Barnsley Area, therefore, mines a whole range of different quality coal (much of it of very good quality) and sells to a variety of different markets (e.g. C.E.G.B., industrial, coking, domestic). So to achieve maximum profitability the Area has to prepare its range of mined coal to suit the available markets.

In the past, small coal preparation plants were sited at individual collieries. To realise the full profit potential of Barnsley Area more preparation plant capacity and flexibility were required. The most economical method of providing this capacity was to construct three major preparation plants to serve the whole Area. Each plant has in fact become the centre of a coal complex - a group of pits, often linked underground, with most of the coal output being raised where the preparation plant is sited, so that it can be fed in directly. The first preparation plant is just being commissioned (at South Kirkby) with the other planned to commence operation in a few years. Thus in the near future Barnsley Area will be divided into 3 coal complexes: the West Side, centred on Woolley; the East Side, centred on South Kirkby; and the South Side centred on Grimethorpe where all coal will be fed through washers and prepared for different markets.

9.2.2 Previous Market Models

The problem of obtaining maximum revenue for a range of coal products with different markets can be easily formulated in linear programming terms. Suppose that there are n different coal products with tonnages $T_1, T_2, \ldots, T_n$ and, for example, ash contents $a_1, a_2, \ldots, a_n$; m different markets of maximum sizes

$S_1, S_2, \ldots, S_m$ and minimum sizes $s_1, s_2, \ldots, s_m$ (i.e. at least s_j tonnes must be supplied to market j). Suppose the maximum acceptable ash content of coal for market j is A_j, $j = 1, 2, \ldots, m$. Denote the price of a coal product, i, in market j by p_{ij}, and the tonnage of each product, i, allocated to a market, j, by x_{ij}. Then the problem of maximizing total revenue is

$$\text{Maximize} \sum_{i=1}^{n} \sum_{j=1}^{m} p_{ij} x_{ij}$$

$$\text{subject to} \sum_{j=1}^{m} x_{ij} \leqslant T_i$$

$$m_j \leqslant \sum_{i=1}^{n} x_{ij} \leqslant M_j$$

$$\frac{\sum_{i=1}^{n} a_i x_{ij}}{\sum_{i=1}^{n} x_{ij}} \leqslant A_j.$$

Note that the ash constraint can be made linear just by multiplying through by the denominator. Similar formulations would be needed for the other quality constraints, such as sulphur content, dirt, moisture, and for size of coal.

Such formulations assume that the pricing formulae for coal are linear, so that the price for a blend of coal is the same as the total price ot its constituents. Such assumptions are found to be true, or at least a useful approximation of the truth. Consequently, it is not surprising that LP was brought to bear on these marketing problems.

The Operational Research Executive first became involved with product-market matching in the 1960's at a colliery level. LP proved to be a useful tool, assisting individual pits in their washing/blending/sweetening/market decisions. Collieries often lacked sufficient capacity to wash all their coal so run of mine coal (i.e. unprepared coal) had to be 'sweetened' with washed coal to produce an acceptable product. Similarly, products had to be blended for particular markets.

These techniques proved to be reasonably successful at a colliery level. Unfortunately, if applied to all collieries in an Area, it could lead to collieries competing for a premium market of insufficient size. It was also often necessary to blend and sweeten coal from different pits, requiring transport, and collieries changing their markets would also upset Area transport arrangements. For these reasons attempts were made to extend the techniques to whole Areas.

The application of linear programming to an Area model proved to be less successful. Whereas the data requirements for a single colliery linear program were just about manageable with the computing facilities available at the time, trying to obtain up-to-date data for all the pits in the Area proved to be almost impossible. Further, much of the data (such as next month's forecast output) were based on assumptions only known at a colliery level. So while the results for a particular pit could be interpreted in the light of assumptions made by colliery management, this same understanding and interpretation was lacking on an Area level. Transport between pits and markets, and between pits themselves, also proved to be a major constraint that was difficult to model.

Gradually, therefore, this work was discontinued. Individual pits tended to specialise in single markets so that linear programming for individual collieries was rarely required. At an Area level the models either failed to provide broadly correct information or proved to be too unwieldy to use. OR work continued, examining particular marketing problems, but no complete model of an Area's coal preparation and marketing functions was developed.

It should be noted that the linear programming market models gradually fell into disuse; there was not a sudden discontinuation. This reflected a gradually growing disullusionment wth the models both by ORE and management, rather than a specific 'failure' of the moels. It is interesting that as linear programs became easier to run, with improved computing facilities and packages, OR scientists began to question their use far more.

9.3 THE CONSTRUCTION OF THE MARKET MODEL

At first sight the request for such a model appeared to be asking for another attempt at the earlier Area models that had fallen into disuse. From a modelling point of view, however, the problem had been greatly simplified by the planned coal complexes. Rather than dealing with eighteen different collieries that are interdependent for blending and sweetening, there are just three independent coal preparation plants, (though these are not independent for marketing considerations). This enormously simplifies the data requirements, transport considerations, and the number of products to be dealt with.

A further advantage conferred on the work by the planned changes in Barnsley Area was that the model was effectively tackling a 'new' problem. The range of coal preparation and marketing decisions that would confront the Area would be much larger than previously encountered, and management had perceived a gap in their existing procedures. It is rare that such gaps occur; if the problem has been around for a long time then management will have evolved a (though not necessarily the best) system to deal with it, and there will be no vacant slot into which a model can be inserted. When new problems occur, however, it is the ideal time to offer OR techniques to be adopted by management, who will consequently not develop alternative methods of their own.

The chances of constructing a successful model therefore looked quite good, and it was decided to pursue the idea.

9.3.1 Specifying the model

Initial discussions with management centred on the following points:

  (i)   Outline of the model.
  (ii)   Optimization v. Calculation.
  (iii) Data Requirements.
  (iv)   Scale of the model.

These points are discussed further.

Outline of Model: The first task was to establish more precisely what model was required. Although it was clear that something was needed, it required more discussion to establish the basic outline of the model. These discussions were mainly with the Deputy Director (administration), the Marketing Manager, Planners, and the Area Coal Preparation Engineer. Also involved were such people as the Area Programmer and Area Chief Scientist. It became clear that the model would have three parts.
  (a) Prediction of output from pits in each complex.
  (b) Calculation of coal available after preparation.
  (c) Allocation of coal to markets.

Optimization v. Calculation: Initially ORE were cautious of an optimizing model, because of the previous difficulties of such models. They suggested a model that would require the specification of coal preparation and market allocation policy, and which would then calculate the resulting revenue. Such a model would have been relatively simple, but Area Management pointed out that this did not meet their needs. With a host of preparation and marketing options available to them, they needed stronger guidance as to what was 'best', what policies would produce maximum revenue. Furthermore, it was necessary to determine the cost of various constraints imposed on the Area. It was therefore agreed that the model would be used for optimizing purposes.

Data for the models: At this stage it was necessary to specify in general the type and amount of data required by the model, to ensure that the Area could obtain the data and would be willing to make this effort. This was agreed.

Scale of the model: All models have an inherent 'scale', i.e. the dimensions (quantities and time) of the model are measured in units that determine the level of detail. It is important to match the scale of the model to the scale that the users work to. In this case it meant deciding the sort of quantities of coal (e.g. tonnes, thousands of tonnes, or millions of tonnes) that the management users dealt with and also the time scale of the decisions (e.g. hours, days, weeks or months) that are made. Once the scale has been decided , this sets the limits of accuracy required by the model and consequently the amount of detail required in the data. It was agreed that the typical timescale of the model would be three months (with a total forward projection of 18 months) and that quantities of coal less than 10,000 tonnes would not be significant.

All of these points required extensive discussion and agreement with management. It should be noted that in all cases it is the management users' needs that are paramount.

9.3.2 Design of the model

With the objectives clear, the model then had to be designed. Two sections of ORE were involved: the section that carried out general OR work for Barnsley Area, and a section that specialises in coal preparation problems. Extensive discussion took place within ORE on the best form of model design, with the consultation of Area management to clarify some points.

A computer model of coal preparation plants ('COALWASH') had already been developed. This would calculate the output from a preparation plant given the run of mine coal input and the specification of the processes within the plant. This model did not, however, automatically specify the best operating policy. Indeed, the best operating policy would depend on the markets available.

Ideally a model was wanted to give the optimal preparation and marketing policy. It seemed likely that the market allocation could be formulated so that linear programming could be applied. The coal preparation process was not so amenable. A representation of the preparation process was searched for that would characterise the operating policy of a particular plant with a few parameters that could be optimized. Such experiments proved fruitless. The COALWASH model would calculate the resultant products given the coal input and operating policy, but there appeared to be no general way of predicting what the best policy would be. It was, however, established that practical difficulties restricted the choice of preparation plant operating policies and that given some idea of the market possibilities, there was only a limited set of 'sensible' operating policies.

In the end it was decided that the model would consist of a suite of computer programs. The first program would predict the coal output from the collieries for each complex over the next 18 months, based on mining plans and scientific data on the amount and quality of coal and dirt in each seam. The second program, based on COALWASH, would determine the products yielded by the coal preparation plant, based on the predictions of the first program and a designated operating policy. The final program would produce an optimal (i.e. generating maximum revenue) allocation of products to markets, given the current prices for coal of different quality in each market. The programs would have to be run several times to try out the range of possible operating policies. The model, as designed, is represented in Fig 9.1. The end result of using this suite of computer programs, which would constitute a model of the real system, would be to give an optimal preparation plant operating policy and market allocation, and the predicted tonnage and revenue with these plans.

It was felt that it was possible to achieve the accuracy required for a model of the agreed scale. The market allocation would be optimized using linear programming. This was not a totally accurate representation since some of the pricing formulae are slightly non-linear but adequate for the circumstances, given that an optimized solution was required.

A full run of the model would probably require several days,

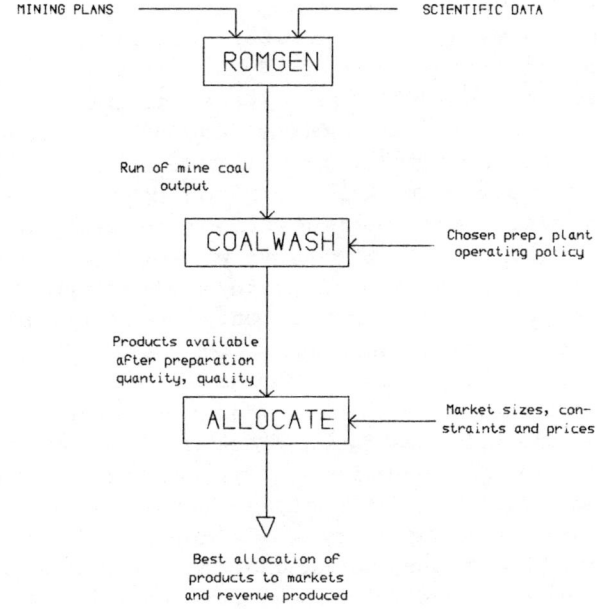

Fig 9.1 Model structure.

for manual examination of the results at each stage, and close co-operation between ORE, Marketing Department and Coal Preparation Branch. The modular design, however, would permit different sections to be run independently. The model could therefore be used for a variety of purposes (e.g. looking at different preparation policies, exploring the effects of changed mining plans, discovering the 'cost' of market constraints) by different departments apart from the main purpose of determining the plant operating and market allocation policies that generates the most revenue.

9.3.3 The pilot model

As it would be some time before all the coal complexes were complete, a full model could not be immediately constructed. Nevertheless, it was desired to obtain some practical experience of using such a model, so that problems could be removed and modifications made: it is often only when users begin to operate a model that necessary extensions and alterations become apparent. It was therefore decided that a pilot model would be constructed which dealt just with the East Side complex, where the preparation plant was already being commissioned. A detailed specification including data requirements was drawn up and agreed with the Area Management.

Each module was written to provide maximum flexibility. This was done so that the modules could be used for a variety of purposes, as mentioned before. The model would then become a flexible tool for Area management (and ORE) and evolve in the direction that proved to be most useful.

The market allocation was optimized using an LP, using a formulation similar to that described in section 9.2.3. A subroutine (written in FORTRAN, as were all the programs) for performing LP's, using a dual simplex algorithm, had already been written by another ORE section. It was established that this was more suitable than using the mathematical packages available, and so it was adapted for use in the market allocation program. Subroutines were written to manipulate the input and output from the LP routine. The input allowed a simple specification of the quantity and quality of each preparation plant product, and the price of each in every market. Market constraints of size and quality were also specified. The quality would be specified by a number of characteristics (e.g. sulphur, moisture, ash) chosen by the user, with no limit on the number or type of characteristics that can be chosen. This 'customizing' of input is a vital feature of models to be used by management; input must be simple, logical and flexible if the programs are to be successfully implemented.

9.4 CONCLUSIONS

Reaction to the initial runs of the pilot model has been very favourable. Area management have shown interest in the results and their validation, and have requested further runs of the model. The initial runs have already identified the correct coal preparation strategy, and have started to explore the possible marketing strategies. Area management are now looking towards the development of a complete Area model embracing the other two coal complexes.

The success of the pilot model results from a number of factors. The cooperation of the Area management has played a significant part. Indeed, it is the users of the model that have provided much of the impetus for its construction. As mentioned earlier, this derives to some extend from tackling a 'new' problem, where there are no established methods: OR models can rush in to fill this vacuum.

Not the least of the reasons for success, though, is that now the physical reality being modelled is far more suitable for linear programming. Although the coal marketing problem has always been mathematically reducible to LP, the solution may be far more easy to derive than it is to implement. Looking at the mathematics alone, it can be easy to forget that dealing with a couple of dozen separate collieries with complicated transport arrangements between them is not the same as controlling a single process such as a refinery, where flows can be stopped, started and adjusted at will. A familiar complaint about the early LP coal market models was that the solution was not "robust". No one wanted to change all the flows in the Area to achieve a better solution as variables changed - hardly surprising given the problems involved.

Although a solution must have some robustness (it would be useless if the solution fluctuated widely when the data were varied within their limits of accuracy), a completely robust solution (one that hardly ever changed), would defeat the purpose of the model. Sensitivity analysis has shown that the solutions obtained from the pilot model are adequately robust, and that the nature of the coal complexes allows products to be blended and re-allocated as the solution requires, at least to within the time scale of the model. The central control inherent in the use of coal complexes allows the use of 'central solutions' derived from LPs.

The aggregating of pits into complexes has also helped in data collection and to smooth out the fluctuations in output of individual pits. Previously, these local variations in pit output could be a major problem, but were too small to be usefully solved by LP. These changes in the Area have produced a reality that can be modelled by LP, and where the solutions can be usefully implemented.

Chapter 10 **Modelling the spread of telecommunications in less developed countries**
G. Walsham

10.1 INTRODUCTION

This chapter is concerned with the role of telecommunications systems in the economic development of less developed countries (LDC's), and with the problems of the managers who control these systems. This is an extremely broad area, and at the present time an adequate and comprehensive framework to structure the problem does not exist. In contrast to the other case studies in this book, we will therefore not be describing a complete model or set of models, but research which has been undertaken to assist in providing such a framework. The eventual aim of this research is to produce quantitative models which can describe the system in as satisfactory a way as, say, a linear programming model can describe the market for coal. The work described here is an early step in this direction.

The next section will describe some of the problems faced by managers of telecommunications systems, in particular in LDC's. The following two sections will describe specific research on the role of the telephone in economic development and on strategic modelling. Finally, some comments and conclusions are presented.

10.2 PROBLEMS OF TELECOMMUNICATIONS STRATEGY

Expenditure on telecommunications, mainly for the telephone service, is a significant part of the public investment programme in most of the less developed countries (LDCs) and this is likely to continue in the foreseeable future. Despite this fact, the links between development goals and telecommunications policy are not clearly understood. For example, it can be argued that the telephone is essentially a luxury good which provides a service to a small minority and therefore telecommunications investment should be given low priority. On the other hand, one can think of the telephone as an important tool in development, enabling wider communication and thus more integrated economic activity throughout the country.

This lack of understanding of the impact of the telephone on economic development is a major problem facing telecommunications managers in the LDC's. In addition to this uncertainty, many of the more routine management problems are especially difficult in the context of a LDC. A number of these are outlined below.

Demand estimation: Estimating the demand for the telephone service is vital both for provision of the service and in the dimensioning of exchanges and the network. There are major problems here for the LDC's, as they are often faced with a low density of demand, giving great uncertainty as to specifically where demand for the service will actually arise. Service also has to be provided to new areas where no direct information about demand is present.

These uncertainties are often combined with high rates of cessation by subscribers, refusals by those on the waiting list when they are offered service, and movement of firms and populations that completely invalidate forecasts. In the longer term, there are uncertainties in the rate of growth, calling rates, call destinations, and in the data on which forecasts are based.

Manpower: It is clear that an adequate supply of manpower is required to install, operate and maintain a successful telecommunications system. Many LDC's face a severe problem in the supply of trained manpower, particularly if they attempt a rapid expansion of the system or if they introduce new technologies without fully anticipating changing manpower needs. Training schools provide part of the answer to these problems, but it is clear that careful planning is needed with respect to the staff of such an institution, its size, the different skills which need to be taught and, perhaps crucially, the responsiveness of the training school to changing needs.

Equipment: The installation, operation and maintenance of equipment probably absorbs a majority of the total planning effort in most administrations, since a multitude of problems exist in this area. Managerial problems can be taken to include the rate of phasing-in of new technology, the compatibility of new technology with old, replacement policy for existing equipment and the difficulties of providing adequate maintenance services. It is clear that equipment policy must be effective at the regional level but must also be integrated at the national level. Account must also be taken of manpower and financial constraints when considering te provision of equipment.

Finance: Constraints in this area include the obvious ones in the LDC's of the availability of capital and shortage of foreign exchange. There is usually the desire to earn at least some given return on capital, and there are often constraints in terms of the desirability of certain policies, for example, marginal cost pricing.

Social: Some of the more indefinite but interesting managerial problems can be brought together under this heading. For example, one could argue that waiting list priorities should be determined by economic need but certain categories of users may well be able to pre-empt such a system by various methods of indirect influence. A policy of subsidising certain subscribers in order to help integrate them into a desirable development pattern may well encounter opposition from those groups who are implicitly required to provide the subsidy. As a final example of a social constraint, some telecommunications administrations are required to employ surplus manpower as a reflection of the desire to impute a lower shadow cost for labour than the salary cost. This problem is not necessarily unique to the LDC's, but the problem is of a greater magnitude in the poorer countries.

Many of these problems can be approached by optimization methods, and have been in other contexts. Linear programming has been applied to manpower planning, as has optimal control theory. Many papers have been written on optimizing the staging of investment in plant. An essential requirement, though is adequate models of the components of the system, and their interactions with each other and the environment in which planning is carried out.

10.3 THE ROLE OF THE TELEPHONE IN ECONOMIC DEVELOPMENT

This section is concerned with some research aimed at obtaining a clearer understanding of the interaction of telecommunications with its environment at a detailed level. To do this effectively, it is necessary to structure a description of the environment which is sufficiently aggregated to produce consistent and coherent results, but incorporates enough detail to expose the mechanisms which the policy makers wish to influence.

Since we are concerned with the impact of telecommunications on economic development, a sensible starting point is to consider the pattern of economic activities in specific places and the pattern of use of telecommunications which this generates. These places will not be simple isolated points. Each place will have economic links with the other places. The nature and strength of these links will depend on the supporting infrastructure, and in particular on telecommunications.

10.3.1 The hierarchy of places

A method is required for classifying places and their links in order to examine the role of telecommunications in the economy. One such method is to rank and classify places on the basis of the degree of economic activity and infrastructure within them. The resulting classification is known as a hierarchy of places. It is then possible to describe a place as, for example, a Rural Centre, and to examine its characteristic links with a neighbouring Urban Centre. These benchmark types of places and links may then be used as a basis for examining other groups of places.

Having established the general concept of hierarchy of places, a further step can be taken by considering natural subsets of the hierarchy which group themselves into distinct regions. Thus research can be concentrated on exploring the relationship between economic development in a particular region and telecommunication use, both internally and externally to the region.

As an illustration of the method of delimiting regions, consider the example of Kenya. The Ministry of Land and Settlement in this country has already carried out a categorisation of places and the levels in the hierarchy are labelled as Principal Towns, Urban Centres, Rural Centres, Market Centres and Local Centres. Thus the boundary of a region may be drawn so as to include, for example, an Urban Centre and the other smaller centres that are associated with it. The region then has a structured hierarchical form reflecting the internal economic links within the region, and the use of telecommunications can be explored in a well-defined context.

Within this framework, specific research on regional development and telephone use has been carried out in Kenya.

This research is the first step towards constructing models of the interaction of telecommunications and regional economic development. Although the research has not yet reached its conclusion, some preliminary results have been generated and a small selection are described below. The analysis was carried out using data on 59 places in the Central, Eastern and Rift Valley Provinces of Kenya. Principal Towns were grouped within Urban Centres, so that four distinct hierarchical levels were considered. Data available for each place included level in the hierarchy, a complete list of subscribers and their main economic activity, revenue per exchange line and the size of the waiting list.

10.3.2 Revenue per line and other factors

A number of relationships have been discovered between revenue per line and other factors.

Level in the hierarchy: A significant relationship was found between revenue per line and level in the hierarchy. In fact, the mean revenue per line for each level in the hierarchy was roughly double that of the next lowest level. This result is not particularly surprising but does support the view that the hierarchy is a useful way of looking at telecommunications usage and that the more intensive use of the telephone at higher levels in the hierarchy is a function of a different profile of economic activities at each level.

Number of exchange line connections: For the set of places at a given level of the hierarchy, a significant correlation was observed between revenue per line and the number of exchange line connections. In other words, calling rates are higher for the larger exchanges and this applies to Urban, Rural and Market Centres. This phenomenon may be due to several factors. A larger exchange offers more opportunity for calling, and this may increase the revenue per line. It may have been noted by the telecommunications administration that a particular exchange generated greater than average revenue, and thus there might have been an expansion there in preference to less profitable exchanges. A final factor may be that there tend to be more exchange lines in the larger places at a particular level in the hierarchy. The place itself may be large because it attracts more economic acitivity than others at its level and this economic activity, in turn, generates more revenue per line. Further research is needed before these effects can be separated.

Availability: At a given place, availibility is defined to be the number of exchange line connections divided by the number of exchange line connections plus the number on the waiting list. Thus an availability of 1 implies that all telephone demand has been satisfied whereas an availability of 0 implies no supply at all.

Using this definition, availability was found to be significantly positively correlated with revenue per line for Rural and Market Centres, but not significantly correlated for Urban Centres. It is sometimes assumed that low availability will cause intense use of the exchange lines that are available. Although different results were observed at different levels of the hierarchy, in no case did low availability result in high use of existing lines. This should not be taken to imply that reduced availability does not cause more intensive use of existing lines but rather, if such an effect exists, that it is probably dominated by the management policy of increasing availability in places with a high revenue per line.

10.3.3 Analysis of economic categories

The set of subscribers in each place were classified into one of 21 economic categories. The number of subscribers in each category was then represented as a percentage of the total number of subscribers. These percentages were used as an input to discriminant analysis to see if the pattern of users of telephones as classified into economic categories was systematically different at each level in the hierarchy. Discriminant functions were developed and tested on a controlled set of places which had been separated for this purpose. A statistical analysis of the results showed that the functions were very good at classifying the hierarchical level of a place on the basis of the economic classification of users of the telephone in that place. The significant variables in discriminating between levels in the hierarchy were plantation and other commercial agriculture, engineering, food processing and textiles, communications, finance and the percentage of public and renters' call boxes.

The same percentages were used as an input to factor analysis to see if there were any underlying regular groupings of telephone users. It was found that four distinct groups of users were present. These groups were labelled as Industry, Agriculture, Commerce and Infrastructure according to the users that were associated with them. As might be expected from the discriminant analysis, the relative predominance of each group was found to be related to the hierarchical level of the place being considered. It was further noted that many of the 21 original categories could be amalgamated by making the merged categories more closely resemble these underlying groups. This last result is important for practical applications of this work since it would not be possible to use a very large number of economic categories for general telephone planning purposes.

The significance of the above analyses is that it is possible to identify a set of relatively few economic uses of the telephone which indicates the level of a place in the hierarchy and thus provides a benchmark for the pattern of usage at each level. Although further research is clearly needed, this could form the basis of a case for giving some users the phone ahead of others at a given level of the hierarchy or of subsidising particular users to come onto the system in order to meet the benchmark pattern in a place where the pattern is deficient.

10.3.4 Further work

The results described above provide some interesting insights into the relationship between telecommunications use and economic development in the selected regions of Kenya. However, the main point of the analysis was to develop methodological concepts, such as those of hierarchies, so as to provide a structure for detailed field research. Following the analysis, a major survey of telephone users was carried out in Kenya in 1979 to collect detailed data on telecommunications use from a sample of subscribers and those on the waiting list. Information was collected on the level and reasons for use, the location and the main economic activities of the user, the main contacts of the user, and user perceptions of the quality of service. Specific information on calls was also collected. Provisional results are now available from this survey and futher analysis is being carried out with the eventual goal being quantitative models of the relationship between economic activity and the use of telecommunications. At the time of writing, first versions of these models have been developed.

10.4 STRATEGIC MODELLING

As a complement to the research on the relationship between economic activity and telecommunications use, work has also been done on the development of strategic planning models. The purpose of these models is to put together a comprehensive description of the telecommunications system viewed from the position of the telecommunications management and thus to enable management to explore future strategies in a structured and quantitative manner. A model of this type has been constructed for a Latin American country and this is now described in more detail.

10.4.1 General structures

For the purpose of the model internal telephone demand is considered in three parts. The first two parts consist of urban and rural exchange connections. The third part consists of public call offices (PCO's), which can be considered in the category of rural telephony in the chosen country since the vast majority of PCO's are planned for the rural areas. The international telephone service is considered separately from the internal service. A simple growth is used for the cost and revenues of the international service and will not be described in detail here.

For each of the three internal services, the model goes through a sequence of calculations for each year under consideration. This sequence is described in outline in Fig 10.1. The solid line gives the actual sequence of the computer program; the dashed line shows other information transfers. It should be noted that the computational sequence can be repeated for any number of years into the future, although estimates will clearly become less reliable for later years and take the form of general indicators rather than forecast values.

10.4.2 Model calibration

Annual data was available on the recent past performance of the telephone system in the chosen country together with forecasts of future performance for the years 1978-82. Data on urban and rural demand, connections achieved and traffic were used to generate demand curves which were a close fit to predicted future values. Other data, including tariff levels and the number of rural PCO's, were read into the model as policy variables, although they were initially set at their past and planned future values. The model generates financial information of operating revenue, split into various categories, operating costs and return on capital. These output values were verified against actual and predicted data.

In summary, a base case of the model was obtained which closely

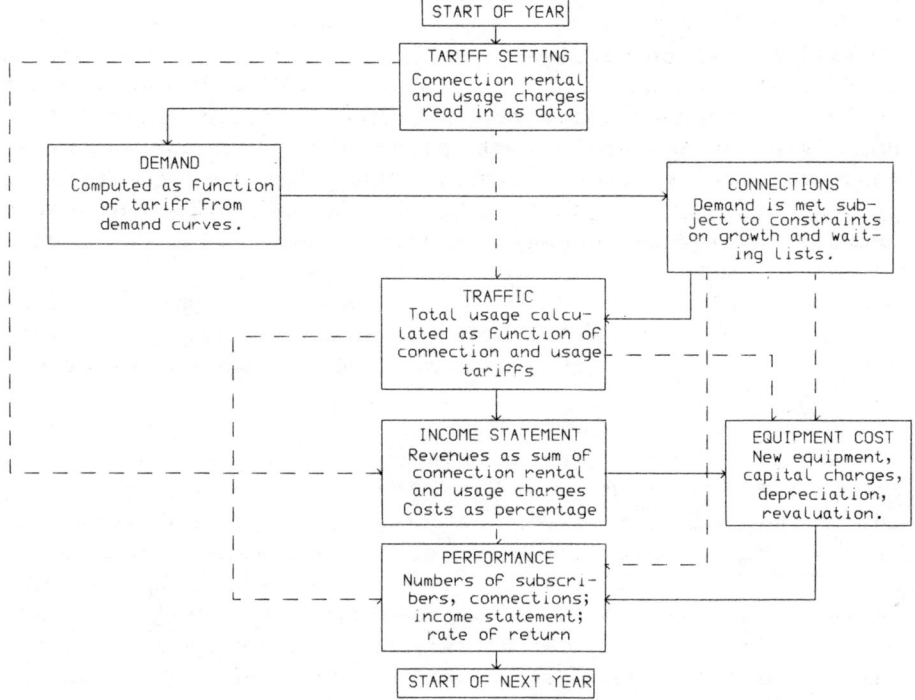

Fig 10.1 Model computational sequence.

matched past and predicted future values. However, some qualification of this statement is necessary. First, the predicted future values may be based on incorrect assumptions. Second, the model was run for the period up to 1987 and no data was available from 1983 onwards. Finally, the one area of model data where no information was available concerned the elasticity of demand for connections and traffic with respect to changes in tariff. Therefore results which are dependent on assumptions in this area need to be viewed with caution.

10.4.3 Results from the model

The application area concerns tariff policy. It has been

suggested that a possible strategy for LDC's is to generate extra revenue from urban services by raising tariffs, and to use this revenue to subsidize rural services. The model has been used to examine some of the implications of this type of strategy.

Two strategies are considered. The first strategy assumes a neutral tariff policy, where prices of all services are raised to keep in line with general inflation rates. The exact values of these rates are not important, but for illustrative purposes general inflation is taken as 8% per annum over the period 1978-1987 and tariffs are raised by 16% every two years, starting in 1979. The second strategy assumes a tariff policy which discriminates in favour of rural services, by raising tariffs on urban services by 18% every two years but on rural services by only 5% every two years. Fig 10.2(a) and 10.2(b) show some results from these alternative strategies.

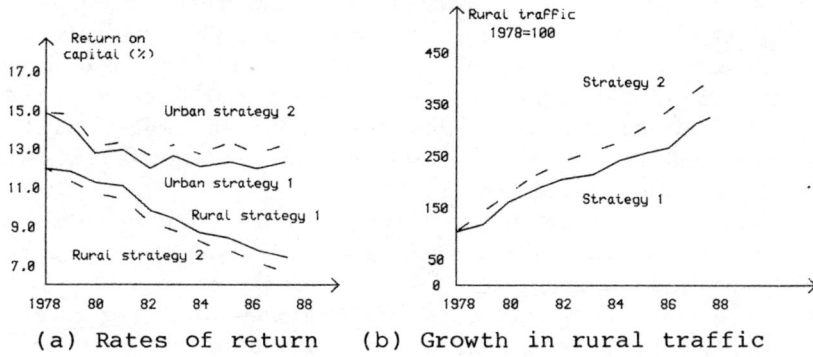

(a) Rates of return   (b) Growth in rural traffic

Fig 10.2 Tariff policy

Fig 10.2(a) quantifies the decline in rural return on capital if the second strategy is adopted. Total return on capital is not given in this figure, but in fact showed a slight increase for the second strategy, indicating that the extra revenue from the urban services more than cancelled out the reduction from rural services. The main advantage of the second strategy is shown in Fig 10.2(b) where rural traffic is 30% higher by 1987. In addition, waiting lists were lower for the second strategy. The only significant disadvantage was a small reduction in the number of urban subscribers - a decrease of 1% by 1987. Of course, there are general economic arguments against the type of cross-subsidy implied by the second strategy but the model results have shown that, from the telecommunications organization's viewpoint, there are significant advantages to the approach particularly if the rural areas are considered to

10.4.4 Further work

A similar model to that described above is being developed for Kenya in order to complement the research on telecommunications use in that country. This latter work has already had considerable impact on the design of the strategic model since, for example, the Kenyan model incorporates the concept of hierarchies in the demand sub-model. At the time of writing, a Kenyan corporate model has been developed and discussed with planners in that country.

10.5 COMMENTS AND CONCLUSIONS

A more detailed description of the aim of research on the role of the telephone in regional economic development and its links with research on strategic modelling is shown in Fig 10.3. The work on the role of the telephone in regional economic development comes on the regional planning side of the diagram. The work has so far concentrated on data collection and the gaining of understanding of the relationships which will be included in models of the interaction of telecommunications with regional economic development.

The work on strategic modelling is shown on the telecommunications planning side of the diagram. The corporate model described in section 10.4 is a first version of a model for strategic planning of telecommunications services as shown in Fig 10.3. Further work is needed to refine the concepts of the model, and to incorporate ideas and structure obtained from the more fundamental research on the role of telecommunications in the economic environment.

Modelling the spread of telecommunications 177

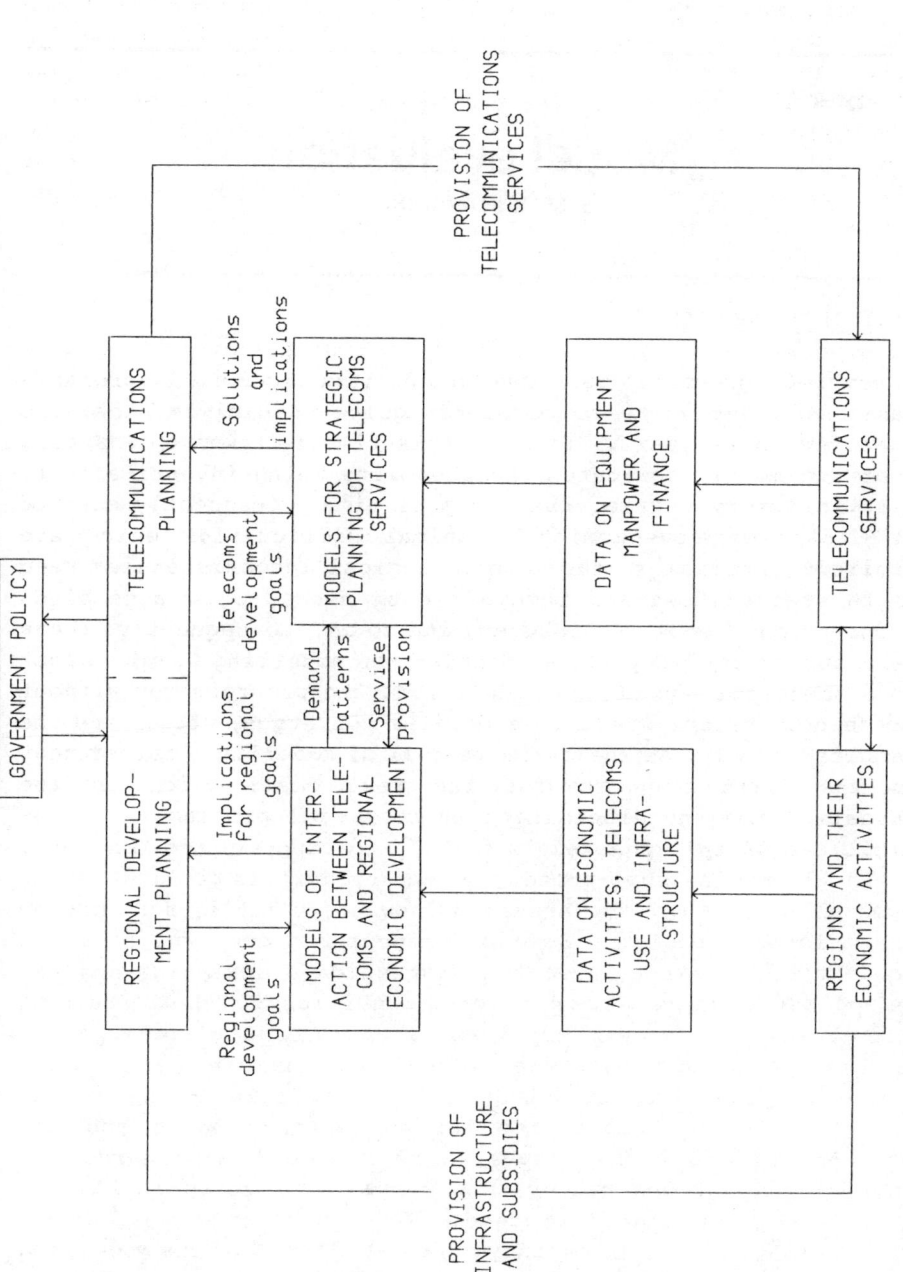

Fig 10.3 Framework for strategic planning

Chapter 11

Model evaluation
J.M. Macieowski

11.1 INTRODUCTION

Developing a model is a long and difficult task. It probably takes at least weeks to complete; quite possibly months, and in a few cases years. Some of this time is taken up ensuring that the model replicates the phenomena being investigated to a satisfactory degree, but a lot of time - frequently most of it - is spent overcoming technical difficulties which are entirely internal to the model. Logical inconsistencies need to be tracked down and removed, data needs to be assembled, program code has to be debugged, and so on. Consequently, there is a strong tendency for a modeller (or modelling team) to look on a succesful run of the model (i.e. the programs run without crashing, the optimization algorithm converges, etc.), and the results from it, as the ultimate goal of modelling, rather than as the first step towards the real goal, which is the investigation and elucidation of some real problem.

The aim of this chapter is to help to counter this tendency. We shall examine some procedures and techniques for evaluating a model - that is, for assessing how useful it is for solving the original problem. Some of these techniques are based on some mathematical (most often statistical) analysis, and can be applied under rather restricted conditions. Others are less formal and more widely applicable, and are justified only by the fact that they have been found to be useful in the past. But it must be remembered that there is infinite scope for the invention of new procedures for evaluating models and the methods which will be discussed here should be regarded as initial suggestions rather than complete prescriptions.

We shall frequently pretend, for convenience that model evaluation is carried out only once, at the end of the modelling exercise. In practice, it will be distributed throughout the modelling process, and will in fact constitute an essential part of it.

11.2 WHAT IS A MODEL?

The other chapters of this book contain examples of very different kinds of model. It may therefore be helpful to have a unified way of thinking about what a model, particularly a mathematical model, is. One way of doing this is to consider every model to be an algorithm for computing

$$o = M(i)$$

where i is the input to the model, and o is its output. M is a function which 'models', or mimics, the 'real world' in certain respects. Evaluating a model means deciding how well this mimicry is performed, on the assumption that i and o correspond to 'real world' variables. Typically, i and o may be system behaviours, in which case M has some 'mechanistic' interpretation. In this case we shall call the model a 'behavioural' model. Alternatively, i may be a set of circumstances confronting a decision maker, and o a decision. In this case M performs some decision-taking role, and we shall call the model a 'decision model'.

Notice that whether a model is a behavioural or a decision model is decided entirely by its intended application, and not by the nature of its input and output. For example, an oil company may use a linear programming model to decide which markets to pursue most vigorously. A second oil company may decide on a policy of concentrating on whichever markets the first company neglects. But to do this effectively it may need to predict which markets these are likely to be. To do this it may use a similar linear programming model to the one used by the first company, with similar input and output. But the first oil company is using the model as a decision model, while the second one is using it as a behavioural model. The evaluations of the two models will consequently be completely different. The first company will judge it by its contribution to its profitability, while the second company will judge it solely by how well it predicts, or is likely to predict, the first company's behaviour. Many economic forecasting models are apparently decision models, but are in fact behavioural. For example, if it is desired to forecast the pattern of energy use in Europe in the year 2000, one way of proceeding is to make assumptions about the costs of various forms of energy supply, the costs of transport from source to user, etc., and then to find that pattern of use which minimizes the total cost to all users.

Another example is the modelling of a human process operator. Here i and o might be time-series of process behaviour and control action. If the modelling is being done in order to simulate the human operator (for instance, as part of a larger simulation) then the model is behavioural. But if it is being done in order to replace a successful human controller by an automatic one, then a decision model is being constructed; in this case what matters is the quality of the resulting control, and not how well the model simulates the human operator.

11.2.1 Behavioural models

The control engineer is used to distinguishing sharply between modelling the system to be controlled and designing the controller. The models we build in this application are behavioural. For example, we might have a model in the familiar form

$$x(k+1) = f(x(k), u(k))$$
$$y(k) = h(x(k)) \tag{11.1}$$

Here we could take

$o = y(k)$,

$i = (x(k-1), u(k-1))$,

$M(.) = h(f(.))$.

Note that this choice is not unique. We could also take

$o = (y(0), \ldots, y(N))$,

$i = (x(0), u(0), \ldots, u(N-1))$

and M the appropriate function constructed from f and h. The former choice implies that we are modelling only what happens at time k, whereas the latter one implies the more usual situation, of modelling over an interval. Note also that i is not just composed of the plant input, since the initial state must be included in it. Different choices of (M,i,o) imply different evaluation procedures, even though we start with the same equations (11.1).

We shall come back to this kind of model later, but for the time being we note that evaluation of behavioural models usually involves comparison with actual data, and that several relatively formal and systematic evaluation procedures exist for doing this, at least for the classes of models most frequently encountered.

11.2.2 Decision models

In operational research and management science the notion of 'model' has a rather different meaning. Here it is usual for the model to include the process of arriving at a decision. Let us consider, for example, the classical diet-mix model of Morris [10].

This model was devised to solve the following problem. A farmer has a herd of cows which need to be fed. Each cow requires a certain minimum amount of each basic nutrient per day, but these nutrients can be obtained from any one foodstuff (grass, grain, etc.) in certain proportions. The price of each foodstuff fluctuates from day to day. Given the ruling prices of foodstuffs, what mixture of them should be given to the cows in order to feed them adequately at least cost? A linear programming model can be formulated as follows. Let there be n foodstuffs, the price of the i'th one being c_i (per kg). Suppose that ω_i kg of it is given to each cow. Then the cost of feeding each cow is

$$c = \sum_{i=1}^{n} c_i \omega_i . \tag{11.2}$$

Let there be m essential nutrients, and suppose that 1kg of the i'th foodstuff contains a_{ij} kg of the j'th nutrient. If each cow needs at least b_j kg of the j'th nutrient per day, then the diet will be adequate if

$$\sum_{i=1}^{n} a_{ji} \omega_i \geqslant b_j, \quad j = 1,2,..,m . \tag{11.3}$$

The minimization of (11.2), subject to the constraints (11.3), is a linear programming problem.

The input i to this model is the set of prices $(c_1,...,c_n)$ and the output o is the set of weights $(\omega_1,...,\omega_n)$. The function

M implemented by the model is the function which maps the set of prices to the set of optimal weights; it depends on the constants a_{ji} and b_j. In this case evaluation is clearly going to be something of a problem; if the model is a good one, then M approximates the 'true' solution, but it is not clear what the 'true' solution is. At least, we cannot directly compare the model output with some set of data. We have to gauge the degree of approximation in informal ways: do the model outputs agree with users' intuitions? Do they agree with decisions taken by successful decision-makers? Have people who have already used the model done well or gone out of business? (Of course, if we can wait for this last piece of information then the evaluation problem largely disappears, for both behavioural and decision models).

Decision models usually include behavioural models within them, often implicitly. In the diet-mix model, there are the linear constraints (11.3) to be satisfied. These arise from a highly nonlinear, 2-state, behavioural model of a cow. Only two states, because the model assumes, in esence, that a cow is either alive or dead; and highly nonlinear, because the cow is assumed to be dead if any

$$\sum_{i=1}^{n} a_{ji} \omega_i$$

is smaller than b_j, alive if each

$$\sum_{i=1}^{n} a_{ji} \omega_i$$

exceeds b_j. Note that this is a steady-state model. It does not describe short-term effects, but only the results of long-term feeding policies. Note also that this inference cannot be made from the mathematical formulation alone, but is in fact an evaluation of the model, based on knowledge of the original problem.

We can consider a decision model as containing a behavioural sub-model and a 'cost' sub-model. The input of the cost sub-model is the input and output of the behavioural sub-model (i. e. a proposed policy and its consequences), and its output is the cost resulting from that policy. We can conceive of a third sub-model, the 'solution' sub-model, whose output is (an approximation to) the solution of any optimization problem inherent in the decision model. This might be a mathematical programming algorithm or an analytic relationship, or it might consist of the informal evaluation by a decision maker of different outcomes associated (via the behavioural model) with different policies. All of these can be inexact to some degree, but we make the simplification of assuming that this sub-model is always correct and exact.

An obvious approach to the evaluation of decision models is to evaluate the two sub-models separately. In the diet-mix model the cost sub-model is probably accurate. The user really does want to minimize the cost of the feed, the cost of each feed is proportional to the amount bought (this may be a local linearisation, but probably an exact one), and the prices are known. On the other hand, the behavioural sub-model is certainly very poor. That is, without very much effort we could probably devise something considerably better than a 'bang-bang' model of a cow, as judged by most evaluation criteria. For example, we could hypothesise a greater range of states of each cow, with the state attained depending on the quality of the diet. But to take advantage of this increased sophistication we would need a more sophisticated cost sub-model, which would take account of the increased profit to be obtained from a better-fed cow. We would now have better sub-models, but paradoxically we now may not have a decision model at all, because we may not be able to find a solution algorithm to use with these more sophisticated sub-models.

The original model, with a rather weak sub-model, may well be the best decision model we can devise. So we see that although it is probably safe to infer that a decision model is good if both its sub-models are good, the converse inference cannot be made. Many decision models - especially linear programming models - owe their success precisely to this fact that, in spite of containing rather primitive sub-models, they 'approximate the true solution' better than any practical alternatives. An example of this has already been discussed in this volume, in chapter 6 describing computer control of a chemical process. Another example is the model used for maintaining water level in canals (chapter 8). More sophisticated models have been proposed for tackling this problem, but they turn out not to be implementable in the application described in the case-study.

11.3 STATIC AND DYNAMIC BEHAVIOURAL MODELS

We divide behavioural models into two types.

11.3.1 Static Behavioural Models

We take 'static models' to mean all those that do not purport to describe the evolution through time of a phenomenon. These are used when investigating 'steady-state' phenomena which either have no variation with time, such as mean values, or equilibrium values of certain stochastic processes, or whose time-variation is deliberately ignored. The use of static models is often due to necessity rather than choice. Queuing theory, for example, deals with the equilibrium properties of queues not because the transient properties are unimportant, but because the analysis of transient properties is too difficult. Much economic theory also deals with static models, assuming, in effect, that dynamic effects within an economy are over within one year, or one quarter, or whatever period is being considered. It is not always possible to overcome the limitations of a static model by abandoning theory and relying on simulation, because the simulation may need to include the optimization of extremely complex problems at each time step, and it may not be feasible to solve these.

Behavioural models are sometimes used purely for the purpose of describing data concisely or conveniently. If it is desired to send someone details of a graph obtained from some experiment, and facsimile facilities are not available, then fitting a polynomial to the graph and sending its coefficients may be more efficient than sending a table of (x,y) pairs. In such a case model evaluation is relatively straightforward. The degree of the polynomial can be increased until it is no longer attractive to send the information in this form.

More often, models are built in order to understand how data were generated, or in order to predict future data. In such cases it is much more difficult to assess the usefulness of a model. A most important consideration is to avoid 'overfitting', that is, the acceptance of a model of unwaranted complexity on the grounds that it gives a very good fit to the data observed so far. The extreme case of overfitting occurs when the postulated model is so complex that it is capable of explaining any set of data that might have been observed. (In the example above this occurs when the degree of the polynomial equals the number of points on the graph). Such a condition is clearly undesirable, because it gives no indication that any pattern or stationarity has been detected in the data; and we need to detect stationarities in order to be able to predict. Putting it another way, if we were estimating parameters in such a complex model using only subsets of the observed data, then the estimated parameters would differ significantly for each subset.

Many formal evaluation procedures for behavioural models involve trading off model complexity against the goodness of fit to the data. One way of doing this explicitly is to treat the model as if it were being used only for data description: compare the amount of information required to transmit the data in the form of the model with the amount required to transmit the raw data. Of course, information must also be transmitted about the nature of the errors in the model, and this introduces the trade-off: the fewer the errors to be described, the less information is needed to do so. Once such a method of comparing a model with the original data has been established, it is possible to use it to compare two rival models of the same set of data; thus it is possible not only to avoid the case of extreme overfitting, but to select the model which is overfitted to the least extent. For specific proposals along these lines see Maciejowski [8], Rissanen [14].

More established techniques for model evaluation introduce the approximation/complexity trade-off less explicitly. Typically, these involve the examination of model residuals, and require the postulation of some probabilistic mechanism which generates these. Increasing the complexity of the model leads to a reduction in the variance of the residuals, but is the reduction large enough to warrant the greater complexity? It is sometimes possible to construct a statistical significance test to answer this question. For details see any good text on mathematical statistics, such as Wilks [15]. Alternatively, it may be possible to estimate the likelihoods of models of varying complexity, and choose as best the one with the highest likelihood.

It should be pointed out that such techniques are only valid under rather special assumptions on the models and the data. Even if one is lucky enough to have most of these hold, one unsatisfiable assumption usually remains. The derivation of test statistics usually relies on asymptotic results that is, results which hold only if infinite amounts of data are available. This is never possible, of course, but the use of these results is safe if enough data is available. The trouble is that one does not known when one has enough, and even if one did, there is usually little possibility of gathering a large amount of data on O.R. and management studies.

An alternative approach to avoiding overfitting is to examine the estimated variance of parameter estimates rather than the variance of residuals. High variance of the parameter estimates implies that the amount of data is not sufficient to allow all the parameters in the model to be estimated reliably, and one ought to look for a more parsimonious model.

Static behavioural models frequently appear in the form of probability distributions, or contain such distributions as key components. The validity of these can be tested by using well-known tests, such as the chi-squared or the Kolmogorov-Smirnov tests (Wilks [15]). Questions of overfitting do not appear to be relevant here, but that is only because it is usual to hypothesise only simple, standard distributions. If one attempted to fit arbitrary distributions to the data, then the danger of overfitting would again arise, and the modeller would have to guard against it.

The main advantage of considering models to be algorithms, rather than sets of equations, is that it implies an assignment of causality. Such an assignment must be made before any decisions can be taken. Thus we do not regard a correlation between rainfall and harvest yield to be a model until it has been decided whether the rainfall is affecting the yield or the yield is affecting the rainfall. Indeed it may be decided that neither is the case, but that both are being affected by some unidentified factor. Such decisions are not always obvious, as is shown by the case-study on modelling the spread of telecommunications (chapter 10). A technique which is sometimes useful for testing hypothesised causality assignments is path analysis (de Neufville and Stafford [12]). This uses the fact that particular patterns of (instantaneous) causality imply particular patterns of correlation coefficients, which can be compared with coefficients estimated from data.

If the assignment of causality reveals a feedback path then one must be extremely careful when interpreting statistical results. This is well known in the (dynamic) system identification literature, but a striking demonstration of what can happen is provided by a static modelling example. If we were trying to determine a policy for sending out fire engines in response to fire alarms, we might try to establish what the (fire damage)/(no. of fire engines) 'transfer function' was. From the data we would find a strong positive correlation between the number of fire engines sent out to a fire and the fire damage. From this we might be driven to conclude that fire engines cause fire damage, and so to ignore all alarm calls. The problem is, of course, that the feedback transfer function has been determined, and not that of the forward path.

Note that there are cases when determination of the forward

path is possible in spite of the presence of feedback. In dynamical linear system identification, for instance, it is possible if the feedback path contains time delays, or is nonlinear, or changes with time (Goodwin and Payne [6]).

Of course, we have not described all, or even most, of the techniques available for evaluating static behavioural models; such techniques occupy much of the statistical literature. But we hope to have illuminated the nature of some of the problems which occur. For an interesting recent discussion of the prolems of evaluating statistical model see Mallows and Walley [9].

11.3.2 Dynamic Behavioural Models

In this section we shall consider the kind of model most familiar to control engineers: a linear, dynamical, input-output (transfer function) or state-space model. We shall not go into details, but limit ourselves to providing signposts to the relevant literature. We shall also not consider models obtained by spectral analysis. The reason for this is that such models are non-parametric and therefore require relatively large amounts of data for successful estimation. As has already been mentioned, O.R. and management science modelling problems are usually characterised by sparsity of data.

Parametric dynamical models take the form of difference equations, whose parameters need to be estimated from data. If the data is univariate and the parameters are assumed to be constant in time, then the selection of a suitable model reduces to the selection of the appropriate order of the difference equation. Such a model can be given the form of a static model by stacking successive inputs and outputs (and states, if appropriate) into vectors, and the parameters can then be estimated using techniques based on those developed for static models. Modifications are required, because the model residuals can no longer be assumed to be independent of each other but several estimation procedures have been developed which bear close resemblances to the classical methods of linear regression, maximum likelihood and Bayesian estimation.

Not surprisingly, then, the model evaluation techniques which are emerging also resemble those techniques used to evaluate static models. For example, Astrom [3] has derived a way of testing whether an increase of order results in a statistically significant reduction in the residual variance. This test is frequently referred to as 'the F-test' in the literature, because it uses the F-distribution.

Another model selection criterion is the celebrated Akaike's

[1] Information Criterion. This is applicable whenever parameters are estimated by a maximum-likelihood method, and their estimates can be assumed to be normally distributed. The criterion states that one should choose that model which minimizes

$$p - \ln(L)$$

where p is the number of parameters estimated, and L is the likelihood function, evaluated at the estimated parameters. This is of very wide applicability, the main restriction being that it is not possible to use maximum likelihood estimation for all types of model. However, maximum-likelihood algorithms exist for the linear dynamic models which are most commonly used. The value of the likelihood function L is estimated from the model residuals, and increases as their variance decreases. Thus Akaike's criterion explicitly trades off model complexity against approximation error.

In recent years other similar criteria have been investigated, all relying on the minimization of

$$N \ln(s) + p \cdot f(N)$$

where N is the number of available observations, s is the sample variance of the residuals and f(N) is some function of N. Attention is usually concentrated on whether such a minimization is consistent in the statistical sense; that is whether it leads to the selection of the correct model as N increases without bound. It appears that good asymptotic properties can be expected if

$$f(N) > \ln(\ln(N))$$

and $f(N)/N \to 0$ as $N \to \infty$. For a detailed survey see Hannan [7].

If multivariate data is being modelled, the problem of model evaluation becomes much more difficult. The essential difficulty is that there is no single measure of 'model complexity'. The number of parameters in a model no longer determines, nor is determined by, the model order. The most complete discussion of the problem is again to be found in Hannan [7]. However, this discussion is technically difficult, and the reader may be better advised to first read Niederlinski and Haydasinski [13], or Young, Jakeman and McMurtrie [16].

11.3.3 Informal techniques

The previous two sections have concentrated on systematic, relatively formal methods of evaluating behavioural models. However, the modeller usually knows a lot more about the system being modelled than he can express in precise mathematical terms. He must therefore use his experience and judgement to decide which, if any, of these techiques are applicable to his problem, and how much weight to attach to the results that they give. His judgement will be helped by the use of less formal procedures, and the following is a check-list of some of the questions that should be considered (in no particular order).

(1) The use of *a priori* information. It is essential that the modeller should use all the knowledge available to him to delimit as far as possible the candidate models. Furthermore, in the absence of specific information he should initially use the most specific rather than the most general model. Otherwise there is the likelihood of overfitting, and the formal evaluation procedures should not be relied on to detect this: they are best used for 'local' decisions, such as 'should the order be increased by one?'.

(2) Cross-check on separate data. The most powerful check of the validity of a model is provided by checking how well it reproduces data which were not used for its construction. For this purpose the collected data is usually split into two subsets, one of which is used for model construction, and the other for model validation. However, this may not be feasible if the original set of data is small. Failure of a cross-check indicates overfitting (by definition), but there is always a great temptation to exonerate the model by appealing to some innate nonstationarity in the system behaviour. This is acceptable only if there is some independent evidence for such nonstationarity; otherwise the apparent nonstationarity may disappear if a more appropriate model is used.

(3) Diagnostic checks. Assumptions such as normality and independence of errors should of course be checked as far as possible. For dynamic models, a large selection of diagnostic checks has been proposed by Box and Jenkins [4]. Graphical checks are particularly useful, and surprisingly often neglected. Plots of the data itself, or of model residuals, or of estimated cross-correlation functions, etc., can immediately make apparent to the eye features which are very

hard to pick out from a column of numbers. A useful suggestion made by Young et al. [16] is that recursive estimation schemes should be used whenever possible, since this enables the convergence of parameter estimates to their final values to be observed. If the estimates converge after only a small part of the data has been processed, and subsequently remain constant, then a cross-check on new data has in effect been accomplished, and one can be fairly certain that the model is not overfitted.

(4) Comparison of various assessments. The result of any model evaluation procedure, whether formal or informal, should be corroborated by at least one other procedure

(5) Consider the purpose of the model. If the model is going to be used to forecast long-term behaviour, then it is not much use to check only the one-step-ahead prediction errors. Conversely, if the ultimate goal is to design a high-gain feedback controller for a system, then only the fast dynamics need to be estimated accurately. If it is desired to predict behaviour during large transients, then data obtained under quiescent conditions may not be useful.

If a macroeconomic model is being used for deciding long-term policy, the possibility that the economic agents will gradually learn what the policy is, and subsequently act so as to confound it (thus completely invalidating the original model) must be considered. On the other hand, if the model is only used to make short-term policy, the learning behaviour may not be important.

11.4 DECISION MODELS

The main problem with evaluating decision models is that one is trying to judge correspondence of the model with intentions rather than with behaviour. As a result, there is even less possibility of 'objective' testing than for behavioural models. Of course the behavioural sub-models of decision models can be tested as discussed previously, but we now wish to concentrate on the remaining aspects.

The intentions which the model must try to match are those of the user: does the cost being minimized resemble the cost which the user would really like to minimize? Does the kind of decision indicated by the model resemble the kind of decision the user really has to make? Do the constraints in the model include constraints which may not be logically necessary, but which the user wants to satisfy? For example, he may wish never to buy less than a certain quantity of materials from a certain supplier, in order to maintain goodwill.

One formal technique which can be a great aid to evaluating these aspects is sensitivity analysis. If we know that both the behavioural and the cost sub-models are rather inaccurate, we can investigate how much we need to change them before a change of policy is implied, or we can investigate how much the policy changes when the sub-models are changed over a region which covers the region of our uncertainty. Unfortunately the Law of Universal Cussedness operates here: the less sensitive is the decision made by the model, then the more obvious is it to the user what the decision should be, and so the less need is there for modelling. Fortunately for us, however, there are real problems which are sufficiently sensitive for their solution not to be obvious, and yet sufficiently insensitive for the solution to be useful.

Sensitivity of the cost using the optimal policy should also be investigated; if this is too high, then it may be better to use a sub-optimal policy whose cost is less sensitive. Astrom [2] gives an interesting example in stochastic control.

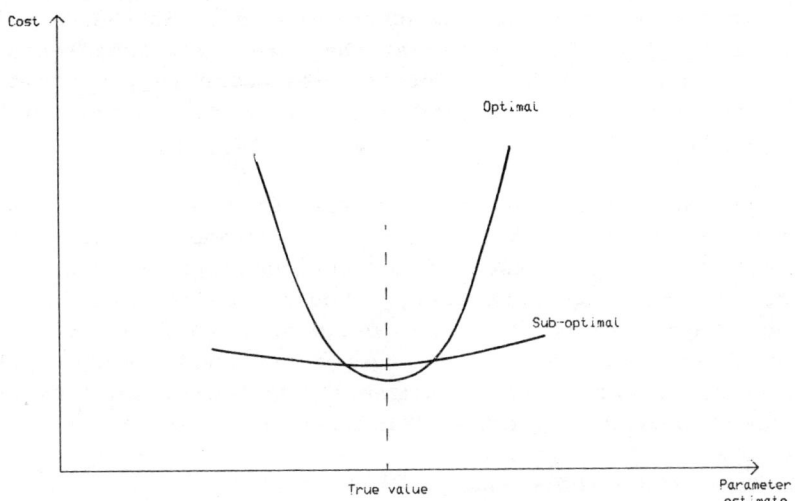

Fig 11.1 Sensitivity to a parameter estimate.

An optimal control is implemented on a system, but its design relies on a system parameter being correctly estimated. The continuous line in Fig 11.1 shows how the cost resulting from

192 Model evaluation

this control law varies with the parameter estimate. The cost is clearly very sensitive to estimation errors. The broken line shows the cost associated with an alternative control law which is sub-optimal because its minimum exceeds the minimum possible. However, the cost for this law is very insensitive to estimation error. If it is expected that the estimate will be erroneous, then the 'sub-optimal' law will in fact be better. It is true that this law can be obtained as the solution to a new optimization problem (in which we minimize, say, the maximum cost), but the need for such a reformulation would not have come to light if the sensitivity of the solution to the original problem had not been considered.

Let us examine some of the problems which arise when attempting to evaluate some of the models appearing elsewhere in this volume.

(1) On-line optimization (Chapter 6)

The systems being modelled here are the most mechanistic of any appearing in the case studies in this book, and although their behaviour is still subject to uncertainty, it is their mechanistic nature which enables very detailed models to be considered. The potential gains from accuracy of representation are very large, so little concession is made in the modelling to analytical tractability. At the same time, implementation of solutions involves continuous data gathering, so that the behavioural model can be kept up to date. The model output is allowed to influence the system directly, by fixing plant set-points.

If we consider either of the cases discussed in this chapter in isolation, it seems that the models stand or fall by their behavioural components, since the cost sub-model, which specifies that profitability should be maximized, seems entirely appropriate for a chemical plant. We should remember, however, that many of the cost components are in fact generated by other optimizing models further up the hierarchy of control, and shortcomings in these models will affect the lower levels.

(2) Control of the economy (Chapter 7)

Here the major problem is validation of the extremely detailed and complex dynamical behavioural sub-model, with very little data being available. There must be a suspicion of overfitting, but this depends crucially on a priori information. Many econometricians argue that there is little doubt about the correct structure of their models. Terms are included in equations because they represent economic mechanisms which are known to exist from experience and theory. If this is correct, then much of validation reduces to technical questions such as the identifiability of parameters in the given structure, and whether the estimation procedures used were appropriate.

The cost sub-model is also rather dubious. A quadratic cost function penalises deviations from a target trajectory symmetrically. This is clearly quite a long way from most people's preferences. For example, high unemployment is usually considered less preferable than low unemployment - a quadratic cost function considers them equally desirable. A sensitivity analysis to changes in the cost function can be performed, and can be useful in spite of our being restricted (by the solution algorithm) to quadratic costs. Suppose that the weighting on deviations from the target trajectory for unemployment is reduced. If the resulting change in the unemployment trajectory tends to be to higher unemployment than before, then the assumption of quadratic costs is not doing much harm. If the change is towards lower unemployment then the quadratic cost is causing a significantly different trajectory to that which would result from using the 'true' cost function. Note that the modelling team had in fact devoted considerable effort to overcoming the defects of the cost sub-model: they developed a 'respecification' procedure which helped them to elicit, from policy-makers, successively more valid cost sub-models.

If an economic model of this kind is just being used to learn about how the economy works - to see what kinds of trade-off are available, for instance - then errors in the sub-models become less important. The policy-maker may find it very useful to see the policies and trajectories obtained from the model, but may then implement a different policy, which he thinks will allow for perceived inaccuracies in the model.

(3) Reservoir control (Chapter 8)

In this case model evaluation is rather straightforward. The behavioural sub-model makes assumptions not so much about the dynamics of the reservoirs, which are just treated as storage tanks, but rather about runoffs and demands. The runoffs to each reservoir are assumed to be in a fixed proportion to one another, certain weeks of every year are assumed to be 'drawdown' or 'refill' weeks, an upper bound for the demand in a given year is assumed to be known at the beginning of the year, and the demand in the refill period is assumed to be small enough to allow the feeder constraints to be disregarded. The plausibility of each of these assumptions is easily verified from historical records.

The objective function which is being maximized is the proportion of demand that can be satisfied. This seems very apt, but only discussion with the user can determine whether it is really what he wants to maximize. The user is probably none too certain himself on this point, and will most likely be amenable to persuasion that this does represent a reasonable objective.

Actually, it is rare for the objective to be self-evident in an optimization problem. Indeed, there is often <u>no</u> objective function. That is to say, there is no strict requirement to maximize or minimize anything, but rather to devise a sensible policy. In such cases, the role of optimization is largely to act as a guide to thinking, forcing the modeller to sharpen his idea of a sensible policy, helping to structure the amorphous space of possible solutions, and narrowing down their often bewildering range. It is for this reason that the solution of very complex problems will often be illuminated by optimizing a simple objective. At the same time, it is important to keep in mind in such a case that the objective being worked with is a surrogate: the narrowing down of possibilities can be detrimental as well as helpful. This last remark applies, perhaps with even greater force, to constraints, which harden when formulated mathematically in a way which is often quite unrealistic.

Returning to the problem of reservoir control, note that the optimization problem is not solved exactly, so the satisfactoriness of the solution has to be checked by simulation. Comparison with the performance of policies followed in the past gives a great deal of confidence in the usefulness of the model. Note that more sophisticated models have been proposed for this problem, but that they turn out not to be implementable in the application tackled in the case-study.

(4) The coal-market model (Chapter 9)

Many operational research scientists are sceptical of optimizing models, and of linear programming in particular. These feelings stem from uncritical application by enthusiastic proponents of these techniques in the early days of their development. This, and the specific difficulties encountered with linear programming market models in the past, made the modellers here initially wary of applying LP. The application itself seems a natural one, provided that the approximations involved are not too great. The great benefit of a linear programming model here is the wealth of information it provides about the optimal solution and its dependence on the parameters of the problem. It follows from this that, given the essential validity of the fundamental linear relationships in the model, the application will stand or fall by how well the final system can communicate that information to the managers who need it. The construction of the linear program is relatively straightforward here, and it is noteworthy that the modelling team put a great deal of effort into ensuring the quality of this communication, both by their involvement of management in the specification and construction of the model, and in the time spent on designing data input and output.

(5) Telecommunications modelling (Chapter 10)

This case study illustrates the very early stages of model building, when there are only partially formed ideas of how the system being modelled behaves, and there is no clear idea as to what the goals of the system should be. Clearly, there is some scope for optimization in parts of the whole telecommunications system, and to some extent on a larger scale, since those who manage the system want to do so as efficiently as possible. Equally clearly, formal optimization is impossible before some sort of behavioural model exists, and the work described in the first part of the study is concerned with first steps toward such a model. Even when such a model exists, it may very well not be sensible to attempt any optimization more formal than that implied in the section on strategic modelling. Here, decision making is improved by providing the decision maker with the means to predict, however approximately, the consequences of different decisions. The value of a model in performing this function lies in the fact that its predictions are at least qualitatively correct, with at least some quantitative accuracy in the shorter term, and partly in the process of education undergone by the user in the course of using the model. Once again, it follows that the quality of the communication between the user and the model is of great importance, as is the user's understanding of how the model works.

11.5 CONCLUSION

Until recently, 'model' meant a scaled-down replica of a physical object or device. Such a model was assessed by the accuracy with which it resembled the appearance of the original object in all respects except scale. In this chapter we have considered models to be defined much more widely, but we have discussed their evaluation essentially from the traditional point of view. We have been concerned with determining how well models resemble 'the real world' in certain respects. But models have other attributes which may make them useful, and which should not be neglected.

Model building imposes a discipline which usually leads to a better understanding of the system being modelled. The value of this may be greater than the value of the model itself - the model may be discarded but the process of constructing it may still be thought to have been worthwhile. Alternatively, attempting to build a model may pinpoint those aspects of the system which are most poorly understood, and on which further investigation should be concentrated.

For interesting discussions of the wider aspects of model validity see Naughton [11], de Neufville and Stafford [12], chapter 12, and the panel discussion on global modelling in Dmowski [5].

REFERENCES

[1] Akaike, H. (1974) A new look at the statistical model identification, IEEE Trans.Auto.Control, AC-19, 716-723.

[2] Astrom, K.J. (1970) Introduction to stochastic control theory. Academic Press, New York.

[3] Astrom, K.J, and Eykhoff, P. (1971) System identification - A survey, Automatica, 7, 123-162.

[4] Box, G.E.P. and Jenkins, G.M. (1970) Time series analysis, Holden-Day, New York.

[5] Dmowski, R.M. (ed) (1974) IFAC/UNESCO Workshop:Systems analysis and modelling approaches in environmental systems, Polish Academy of Sciences, Warsaw.

[6] Goodwin, G. C. and Payne R.L. (1977) Dynamic system identification; experiment design and data analysis, Academic Press, New York.

[7] Hannan, E.J. (1980) System identification, in Proc. NATO Advanced Study Inst. on Stochastic Systems, Les Arcs, Hazewinkel,M. and Willems, J.C. eds.

[8] Maciejowski, J. M. (1979) Model discrimination using an algorithmic information criterion, Automatica, 15, 579-593.

[9] Mallows, C.L. and Whalley, P. (1980) A theory of data analysis?, Bell Labs Technical Memorandum, 80-1215-7.

[10] Morris, D. Using linear models:formulation, optimization and interpretation, The Open University Press, Milton Keynes.

[11] Naughton, J. (1975) Scientific method and systems modelling, The Open University Press, Milton Keynes.

[12] de Neufville, R. and Stafford, J. H. (1971) Systems analysis for engineers and managers, McGraw-Hill, New York..

[13] Niederlinski, A. and Hajdasinski, A. (1979) Multivariable system identification - a survey, Proc. 5th IFAC Symp. on Identification & System Parameter Estimation, Darmstadt, pp43-76, (to be published by Pergamon Press).

[14]   Rissanen, J. (1978)   Modeling by shortest data description, Automatica, 14, 465-471.

[15]   Wilks, S. S. (1962) Mathematical statistics, Wiley, New York.

[16] Young, P.C., Jakeman, A. and McMurtrie, R. (1980)   An instrumental variable method for model order identification, Automatica,16.

Index

active set 4
  strategy 30
adjoint
  equations 46
  variables 46
admissible
  directions 22
  functions 39
algorithms
  for constrained
      optimization 21
    dual methods 21,23
    primal methods 21
    penalty methods 23
    barrier methods 27
    augmented Lagrang-
      ian methods 27
  for unconstrained
      optimization 11
    conjugate gradient 16
    quasi-Newton 18
    requiring derivatives 13
    requiring second
      derivatives 21
    without derivatives 13
applications
  economic models 123
  industrial power
      station 111
  market model 156
  olefine plant 105
  reservoir control 141
  telecommunications
      in LDC's 166

backward shift operator 127
Balakrishnan's method 54
barrier method 27
basic solution 78

basic optimal solution,
      existence of 85
basis 79
  initial choice of 86
block angular structure 98
brachistochrone 37,41
British Waterways Board 141

calculus of variations 36
  extensions to simplest
      problem 43
  necessary conditions 40
  simplest problem 39
causality 186
control of industrial
      process 105
complementary slackness 91
conditioning 14
condition number 14
conjugate
  directions 12
  gradient algorithms 13,16
constraints 1
  active set of 4
  constraint qualification 6
coordination 63
  dual 66,69
  primal 66
co-state
  equations 46
  variables 46

Dantzig-Wolfe de-
      composition 98
  algorithm 102
  column generation 101
  complete master problem 99
decentralized system 62

decision model 181
decomposition 61
  of linear program 97
diet-mix model 181
discretization of optimal
  control problems 51
discriminant analysis 171
drawdown time 146
dual
  coordination 69
    economic inter-
      pretation 70
  algorithms 21,23
  problem 70,89
dynamic
  optimization 26
  programming 47,117

Euler equations 39
  first integral of 41

Farkas' lemma 3,5
feasible
  solution 2
first variation of a
  functional 40
F-test 187
functional 36

gradient
  conjugate 13,16
  methods in optimal
    control 72

Hamiltonian 46
Hamilton-Jacobi equation 48
hierarchical
  optimization
  system 62
hierarchy of places 169

I.C.I. 105
information criterion 188
integer programming 31
isoperimetric problems 38,44
iterative respecification 132

Lagrange multipliers 2
  economic interpretation 8
  in linear programming 89

Lagrange multiplier
    functions 45
Lagrangian
  function 2
linear program 31,77
  basic solutions 78
  basis 79
  complementary slackness 91
  decomposition 97
  duality 89
  form of 77
  post-optimality analysis 87
  revised simplex
    algorithm 80
linear-quadratic
    control problem 48
LQG problem in econo-
    metrics 127
line search 12,110

macroeconomic models 124
  size of 125
mathematical programming 1
matrix generator 94
maximum demand tariff 115
minimum principle 45
model evaluation 178
models as algorithms 179

National Coal Board 156
non-differentiable
    programming 31

objective function 1
  quadratic 11
optimal control 44
  of reservoir systems 143
optimality conditions
  first order 5
  second order 8
optimal value function 47
overfitting 184

path analysis 186
parametric
  decomposition 66
  linear programming 88
penalty function 23,54
perturbation model 129
pivot step in linear
    programming 82

Poisson equation	44	sensitivity analysis	191
Pontryagin	47	sequential unconstrained minimization	23
post-optimality analysis	87	shadow prices	8
price coordination	72	simplex method	77
principle of optimality	47,117, 145	simplex multipliers	82,93
		sparsity in quasi-Newton methods	20
		state constraints	56,145
quadratic		steepest descent	
objective	11	direction	12
programming	28	algorithm	22
termination property	12	sub-models	183
quasi-newton			
algorithms	18		
equation	18	tariff policy	174
		telecommunications	
		systems	166
real-time optimization	105	strategy	172
recursive			
estimation	190		
quadratic programming	28		
report writer	94	unconstrained minimization	11
reservoir control	141		
on drawdown	145		
on refill	147		
restarting in conjugate gradient methods	17	variable end point problems	32
revised simplex method	80		
convergence of	84		
example	82		
Ricatti equation	49	weak duality	90
Ritz method	52		
robustness	165		
role of telephone in development	168	zigzagging	14